A NATURALIST'S GUIDE TO THE

BUTTERFLIES & MOTHS OF AOTEAROA NEW ZEALAND

Carey Knox

JOHN BEAUFOY PUBLISHING

First published in the United Kingdom and Aotearoa New Zealand
in 2024 by John Beaufoy Publishing Ltd
11 Blenheim Court, 316 Woodstock Road, Oxford OX2 7NS, England
www.johnbeaufoy.com

Copyright © 2024 John Beaufoy Publishing Limited
Copyright in text © 2024 Carey Knox
Copyright in photographs © Carey Knox unless otherwise specified
Copyright in illustrations © 2024 Carey Knox
Copyright in maps © 2024 John Beaufoy Publishing Limited

Photo Credits
Front cover: *main image* Central Otago Copper; *bottom left* Red Admiral, *bottom centre* Dotted Green Carpet, *bottom right* Orange Astelia Wainscot, all © Carey Knox
Back cover: Pug Moth (*Pacifphila punicea*) © Carey Knox
Title page: Green Coprosma Carpet © Carey Knox
Contents page: Tutu Green Spindle © Carey Knox

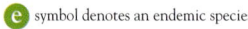 symbol denotes an endemic species

All rights reserved. No part of this publication may be reproduced, stored in a retrieval system or transmitted in any form or by any means, electronic, mechanical, photocopying, recording or otherwise, without the prior written permission of the publishers.

Great care has been taken to maintain the accuracy of the information contained in this work. However, neither the publishers nor the authors can be held responsible for any consequences arising from the use of the information contained therein.

ISBN 978-1-913679-66-8

Edited by Krystyna Mayer
Designed by Nigel Partridge

Printed and bound in Malaysia by Times Offset (M) Sdn. Bhd.

· CONTENTS ·

Introduction 4

Geography, Climate & Habitats 5

Butterfly or Moth 7

Butterfly & Moth Anatomy 7

Life Cycle 9

Conservation 10

Finding & Studying Moths & Butterflies 11

Identifying Moths & Butterflies 11

How to Use This Book 13

Glossary 14

Species Descriptions 17

Further Reading 172

Index 173

Acknowledgements 176

INTRODUCTION

Aotearoa New Zealand is a temperate utopia in the Pacific Ocean, which holds some of the most exceptional flora and fauna on the planet. Long isolated since its split from the ancient super-continent Gondwana, about 80 million years ago, and now nestled between the temperate latitudes of 35° S and 47° S, remarkable animals, such as moas (order Dinornithiformes), kiwis (order Apterygiformes) and the Tuatara *Sphenodon punctatus* developed in this unusual, remote land. Diversity and uniqueness flourished concealed from the rest of the world.

Much like Aotearoa's birds and reptiles, the moths and butterflies (Lepidoptera) of Aotearoa developed along a unique pathway. Some more or less globally distributed Lepidoptera families, such as tussock moths (Lymantriinae) and tiger moths (Arctiinae) are completely absent or under-represented in Aotearoa. Also, in common with many islands, there are few butterflies (less than 50 species). Other families are unusually diverse, such as the litter-feeding Oecophoridae moths (more than 150 species), Tineidae (about 70 species), and the largely moss-feeding Scopariinae with more than 130 species. Aotearoa is also known for the presence of ancient, relict groups, such as the Micropterigidae with 19 species, and Aotearoa's only described endemic family, Mnesarchaeidae, with 14 species.

An astonishing and unique array of moths and butterflies call Aotearoa home, with more than 2,000 species thought to be present. This makes the Lepidoptera the country's third largest order of insects after beetles (Coleoptera) and flies (Diptera). Roughly 1,750 Lepidoptera species have been scientifically described, with an additional 300 or more awaiting scientific description. Many undescribed species exist due to the small number of professional lepidopterists in Aotearoa, who are tasked with working on an immense number of species. Describing a new species is a complicated process that can take several years. A combination of field data and observations, microscope work, DNA research, detailed descriptions of specimens in collections and/or the rearing of caterpillars may all be required. However, with time, progressively more species are being scientifically described.

More than 90 per cent of the country's Lepidoptera are endemic, and thus found nowhere else in the world. This is the highest proportion of endemic butterflies and

The Bright Green Carpet Asaphodes philpotti *is endemic to Aotearoa.*

The Cinnabar Moth Tyria jacobaeae *was deliberately introduced into Aotearoa to control Ragwort.*

■ Geography, Climate & Habitats ■

moths in any country in the world, making Aotearoa a unique and special place to study Lepidoptera. About 160 species have also been introduced, some deliberately and some accidentally, since European settlement. Furthermore, about 30 migrants or vagrants occasionally turn up in Aotearoa, usually after making the journey across the Tasman Sea from Australia.

Geography, Climate & Habitats

Aotearoa New Zealand comprises three major islands: Te Ika-a-Māui/North Island, Te Waipounamu/South Island and Rakiura/Stewart Island. There are also numerous smaller islands surrounding parts of the country, including the Chatham Islands located far east of the South Island and the subantarctic Snares, Antipodes, Auckland and Campbell Islands far to the south. The country is located on a major plate boundary. Over millions of years this has intensified tectonic uplift, volcanic activity and mountain building. Volcanoes are concentrated in the Central North Island and Auckland Region, where the Pacific Plate is actively being subducted below the Indo-Australian plate. In southern Aotearoa, the Indo-Australian plate is being subducted under the Pacific Plate, which has resulted in extensive mountain formation. Today, hundreds of tall mountains are strung along the spine of the South Island, known as the Southern Alps, including Aotearoa's highest peak, Aoraki/Mount Cook (3,724m).

Geological activity has also resulted in extended periods of glaciation, which, in combination with mountain building, has promoted speciation in many groups of flora and fauna. The Lepidoptera have evolved to take advantage of a wide range of habitats created by the country's tumultuous geological past, including coastlines, wetlands, herb fields, saltpans, shrubland, forests, subalpine shrubland, tussock grassland, and high alpine areas to well in excess of 2,000m above sea level. Aotearoa has four distinct seasons, with a mean annual temperature of 10° C in the South Island and 16° C in the North Island. Parts of the country can experience 35° C heat in summer and extreme cold in winter (> –15° C). This 50° C temperature range can be experienced within the space of a year, in places like Central Otago in the South Island. Other, more coastal or North Island localities, experience substantially less variation in temperatures across a year. High alpine zones can be covered in snow for more than half the year, so Lepidoptera in these locations must make the most of the short summer season.

Some regions of Aotearoa are very wet and others very dry. For example, Fiordland has a yearly average of about 7m (7,000mm) of rain or 200 rain days, whereas Central Otago is the driest region of Aotearoa, receiving less than 400mm of rainfall annually. This substantial variation in geology, habitat, elevation and climate across the country provides a wide range of microhabitats that have been fully exploited by its endemic Lepidoptera, and promoted speciation and diversity across millions of years. More recently, since European settlement (in the last 200 years or so), some introduced Lepidoptera species have also become established.

Lepidoptera are spread across the entirety of Aotearoa, but generally more pristine and botanically diverse environments hold a greater number of species. However, even

◾ Geography, Climate & Habitats ◾

in the most densely populated cities, Lepidoptera can be found. Some species, such as *Epyaxa lucidata*, the Common Forest Looper *Pseudocoremia suavis*, Slender Owlet Moth *Rhapsa scotosialis* and Aotearoa Cutworm *Ichneutica mutans*, seem to do reasonably well in urban environments (or suburbia). In fact, more than 400 species (close to 25 per cent of Aotearoa's described Lepidoptera) have been recorded in the heart of Auckland City (between Manurewa and Albany), according to the website iNaturalist (p. 173).

The introduced or cosmopolitan moths, such as the Tropical Armyworm Moth *Spodoptera litura*, Apple Looper *Phrissogonus laticostata*, Dark Sword-grass *Agrotis ipsilon* and Australian Bag Moth *Cebysa leucotelus*, seem to do particularly well in urban areas. Cities or towns with an abundance of native plants in parks, roadsides (or islands), shared community spaces and people's gardens will generally have a reasonable diversity of endemic Lepidoptera. Sometimes the diversity of moths in towns and cities can be greater than in adjacent rural areas, particularly if these rural areas are intensively farmed and have little native vegetation. This is due to the greater number and diversity of host plants available to the Lepidoptera. Regardless of where you live in Aotearoa, though, looking at moths on your house windows, under your outside lights or in your garden can be a fruitful and enjoyable pastime.

Often, venturing out of the city will reward you with a greater diversity of moths. Certainly, a greater proportion of endemic moths, and often rarer species, can be seen outside the city limits. This is particularly so in protected natural habitats along rugged coastlines, pristine native forests, diverse mature shrubland, tussock grassland, wetlands, saltpans and alpine zones. The diversity of moths is high in both northern and southern regions. For example, close to 800 species have been recorded in the Auckland region (including 650 species in the Waitakere Ranges alone), and the Otago region is known to hold more than 1,000 described species of Lepidoptera. Diversity can also be high in both damp forested areas and drier regions. For instance, Fiordland National Park holds more than 700 species, including many beautiful rarities, such as the Exquisite Carpet *Asaphodes*

Rocky subalpine habitat above beech forest in Northern Fiordland. A diverse range of shrubs, grasses, tussocks and herbs is present, creating abundant microhabitats for Lepidoptera.

Tussock grassland in the Lammermoor Range of eastern Otago. A diverse tussock grassland contains flax, Dracophyllum shrubs, ferns, subshrubs and a beech-forest remnant.

Butterfly or Moth

adonis and the rare spindle moth *Tatosoma monoviridisata*. The markedly drier Central Otago holds a similar number of species, emphasizing the adaptability and range of habitats used by the country's Lepidoptera.

Butterfly or Moth

There are no major differences between butterflies and moths – they are both common names given to groups of insects within the order Lepidoptera (Greek for 'scaly wings'). Most butterflies and moths have a similar life cycle. Butterflies are nestled within the moth family tree and can be considered a subgroup of 'modified day-flying moths'.

It is important to recognize that the fact that a Lepidoptera species is brightly coloured and flies by day does not necessarily mean that it is a butterfly. In fact, Aotearoa has a diverse array of beautiful and brightly coloured day-flying moths. The best way to distinguish a butterfly from a moth in Aotearoa is to look for the slightly swollen or club-shaped tips that are present only on the end of the antennae in butterflies and not in moths (which have feathery or thread-like antennae without swollen/rounded tips). Furthermore, often in butterflies the forewings sit further forwards than in moths when the insect is at rest.

Butterfly & Moth Anatomy

The bodies of moths and butterflies, like those of other insects, are divided into three regions, the head, thorax and abdomen. The head has a pair of compound eyes. The antennae, which detect important chemical cues such as pheromones from the opposite sex and volatiles from their host plants, are placed just above the compound eyes. Antennae are segmented and in male moths often bear lateral branches called pectinations. Most Lepidoptera have a proboscis (tongue) and a more or less prominent pair of labial palps

Not to be confused with a butterfly, the Western Tiger Moth Metacrias erichrysa *is a brightly coloured endemic day-flying moth.*

A butterfly in the genus Lycaena. *Note the slightly swollen tips on the ends of the antennae that help distinguish it from a day-flying moth.*

■ Butterfly & Moth Anatomy ■

Adult moth and butterfly anatomy illustrated on the Rusty Hebe Looper Dasyuris callicrena.

- Antenna
- Forewing costa
- Apex
- Compound eye
- Thorax
- Forewing termen
- Labial palp
- Hindwing
- Proboscis
- Foreleg
- Abdomen
- Tornus

Adult moth anatomy illustrated on Meterana asterope.

- Antenna
- Collar
- Basal line
- Dorsum
- Antemedian line
- Claviform spot
- Orbicular spot
- Median line
- Reniform spot
- Postmedian line
- Subterminal line
- Terminal line
- Apex
- Fringe
- Tornus
- Terminal area

LIFE CYCLE

that project from below the head. The proboscis is extensible and often coiled up when not in use. It forms a tube that the butterfly or moth uses to draw up food such as nectar. Some moths, however, do not have a proboscis, and do not feed as adults, for example the ghost moths (Hepialidae). The most primitive moth family, the Micropterigidae, has chewing jaws and feeds on pollen and fern spores.

The thorax consists of three sections: protothorax, mesothorax and metathorax. Each of these has a pair of legs and the latter two have a pair of membranous wings clothed in scales. The wings are structurally supported by tubular veins. Wing venation varies between groups of moths and can be useful in classification. The leading edge of the wing is termed the costa and the outer margin (furthest from the head) is the termen. The trailing edge is termed the dorsum, and the 'corner' of the wing where dorsum and termen meet is the tornus. Other features used to help describe moths include the various lines on the forewing (such as postmedian line and terminal line), or spots such as the reniform spot and orbicular spot (see diagrams opposite).

The abdomen contains about 10 segments with the last 2–3 modified into the complex genitalia of each species. The male abdomen ends in a pair of valvae used to hold the female during copulation. The end of the female's abdomen has an ovipositor for depositing the eggs.

LIFE CYCLE

Both butterflies and moths have three immature life stages before emerging as adults: egg, larva (caterpillar) and pupa. Timing of the life cycle varies enormously between species and habitats. Any of the immature phases can pass through a resting phase, particularly over the winter months, when development is effectively paused (termed diapause). The adults lay their eggs on plants or other surfaces such as rocks or lichens. Larvae (or caterpillars) hatch from the eggs, feed and grow. As they grow, they moult (shed their skin), usually four times, and the period between moults is referred to as an instar (for example, the third instar occurs after two moults). Larvae are usually very well camouflaged on their host plant in shades of green or brown, but in some cases species that are distasteful to predators

Caterpillars of Meterana ochthistis *(left) and* Metacrias huttonii *(right)*

have bright warning colouration to try and dissuade birds, reptiles, amphibians, mammals or insects from eating them. Many larvae feed at night. When fully fed, they pupate in the host plant, soil or leaf litter. Most moth larvae pupate inside a silken cocoon, but those of butterflies and a few moths (for example many Elachistidae and Pterophoridae) are attached to the substrate only by a small pad or girdle of silk. Inside the pupa the caterpillar body must be broken down and completely re-formed into the winged adult before emergence.

The caterpillars of Aotearoa's butterflies and moths can be split loosely into three ecological groups: those feeding on leaf litter and dead wood and fungi (including lichens), those feeding on non-vascular plants, such as mosses and liverworts, and those feeding on vascular plants (such as grasses, tussocks, shrubs, vines and trees). Species also vary enormously in their selectivity of host plants, often termed as either monophagous or polyphagous. Those like the Forest Semilooper *Declana floccosa* are polyphagous (in terms of host plants), meaning that the larvae feed on a wide range of native and exotic shrubs, vines and trees. Exotic hosts for the Forest Semilooper include pines (*Pinus* species), Douglas Fir *Pseudotsuga menziesii*, larches (*Larix*) and eucalyptus. Native hosts include *Muehlenbeckia australis*. In contrast, species such as *Nola parvitis* (the only native representative of the global family Nolidae) only feed on *Helichrysum lanceolatum* shrubs, so are solely reliant on that plant for survival. Similarly, the stunning Grand Olearia Owlet *Meterana grandiosa* only feeds on a small range of small-leaved divaricating *Olearia* shrubs (or tree daisies), such as *O. odorata* and *O. hectorii*.

CONSERVATION

Endemic species that have limited distributions and/or are monophagous (restricted to one or a few related host species) are generally very vulnerable to population decline or extinction. This is because, if one particular shrub or tree that is vital to them as a host plant is lost due to forest or shrubland clearance, the moth or butterfly will be lost with it. For example, both the Grand Olearia Owlet *Meterana grandiosa* and Exquisite Olearia Owlet *M. exquisita* have declined in extent and overall numbers in recent decades due to the clearance of shrubland containing their host plants, small-leaved *Olearia* shrubs. In contrast, a polyphagous species may simply be able to adapt to using another shrub or tree species in the vicinity. Due to this, an understanding of rare Lepidoptera in each region of Aotearoa, and preservation and planting of their favoured host plants, is critical for conservation.

Some key plants for the country's Lepidoptera that host a variety of moth species include: *Muehlenbeckia* species (vines or small shrubs such as the Scrambling Pohuehue *M. complexa*, native hebe/veronica shrubs such as the Boxleaf Hebe, *Veronica odora*, *Urtica* species (nettles), *Olearia* shrubs or trees (daisy bushes), *Melicytus* (shrubs or trees), *Ozothamnus* (shrubs), *Celmisia* (mountain daisies), *Pimelea* (rice flowers), *Kelleria* (miniature shrubs), *Cardamine* (bittercresses and toothworts), *Coprosma* (shrubs or small trees such as Mingimingi *Coprosma propinqua*), *Pseudopanax* (lancewoods and five fingers), *Dracophyllum* (shrubs and trees), *Gaultheria* (wintergreens), *Carmichaelia* (brooms), *Sophora*

Finding & Studying

(kowhai trees), *Epilobium* (willowherbs), *Ranunculus* (buttercups), *Clematis* (vines), *Carex* (sedges), *Poa* (tussocks and grasses) and *Chionochloa* (tussocks).

Planting of the above native plants, as well as other natives, is recommended to provide key habitat across Aotearoa so that the beautiful butterfly and moth species highlighted in this book (and numerous others) can flourish and remain for future generations to study and admire.

Finding & Studying Moths & Butterflies

Observing and studying Lepidoptera can vary from a casual hobby through to a professional career. The great news is that it is very simple to get started. Many enthusiasts start by simply identifying butterflies that flutter through the garden by day, or what moths occur on their house windows at night. A camera with a macro function or lens is useful for providing a photographic record of the butterflies or moths you find. To observe a greater number and diversity of moths at night, consider investing in a better light source to attract them. UV lights of 10–40 watts are very effective (moths are strongly attracted to UV or actinic light) and are inexpensive to buy, so it need not be an expensive hobby. Virtually anyone can take part.

Another strategy for locating moths at night is to simply walk through forest or shrubland with a torch and camera and see what moths take to the air or emerge from disturbed vegetation as you pass by trees or shrubs. Gently tapping vegetation such as ferns, shrubs or trees may cause moths to take flight and thus be seen.

If late nights do not appeal, it is also enjoyable to head out during the day to see what day-flying species of Lepidoptera can be found. Walking through grassland, shrubland, wetlands or forest in sunny or overcast weather can be useful in seeing what species can be found. Gentle 'tapping' of tussocks, shrubs, trees, clumps of ferns, or tree-fern skirts may cause diurnal species or resting nocturnal species to take to the air. For the more adventurous, heading into the mountains and looking for day-flying species above the tree line can be very rewarding. In the alpine zone brightly coloured day-flying moth species in the genera *Notoreas*, *Aponotoreas*, *Paranotoreas*, *Dasyuris*, *Asaphodes*, *Arctesthes* and numerous others can be found fluttering over tussock grassland, wetlands, cushion fields, shrubland and herb fields. Keep an eye out for fast-flying species in the South Island mountains, such as tiger moths *Metacrias*, and the Streaked Inanga Looper *Declana glacialis*. There are also stunning butterflies to look out for above the tree line, such as the black mountain ringlets *Percnodaimon*, copper butterflies *Lycaena*, tussock ringlets *Argyrophenga*, and Butler's Ringlet *Erebiola butleri*.

All moths and butterflies illustrated in this book were photographed in the wild. Some were photographed in situ on rocks, trees, grasses or shrubs. Others were captured in a hand net or in small plastic bottles and carefully transferred onto a sheet of white paper to be photographed. Every endeavour was made to not harm the species and to release them alive back to where they were found soon after photography.

Identifying Moths & Butterflies

When identifying Lepidoptera start big and work your way down. Ask yourself the following questions: is it a moth or butterfly? What is the location? What habitat was it found in? Is it flying by day or at night? What colour is it? Is it large, medium or small? What is the shape of the forewings and hindwings? Are there any distinctive lines, patterns or markings on the forewings and/or hindwings? Considering these questions may help you pin down what family or genus it belongs to.

Refer to this book and others, and articles on Aotearoa New Zealand's Lepidoptera, to see if you can find a similar match, keeping in mind that some species can be variable in appearance and may look quite different from the images provided in this (or other) books or online resources. Also, occasionally males and females of the same species can look quite different from each other (termed sexual dimorphism). Remember to look at the antennae to help differentiate between male and female moths (males often have obvious pectinations on the antennae). Perhaps most importantly, take a photograph to create a record of what was seen while the insect is around. Even a mobile phone picture can be sufficient, at least for most of the medium–larger Lepidoptera, but a camera with a macro lens may be more useful for smaller moths. Try and obtain photographs from several angles to show features of the forewings, hindwings, head and even the underside (where possible).

This book can be used to identify the majority of common species you may come across and a few rarities, but a further great resource for identifying butterflies and moths is the website/app iNaturalist (www.inaturalist.nz). Simply create a profile and upload photographs of your butterfly and moth observations. Experts are available on the site to confirm (or refute) your identifications and help you with any rarer Lepidoptera species you may come across. Keep in mind that for some genera it is difficult to pinpoint exact species from photographs, so the identification may not be able to proceed past genus level. With time, you will learn to recognize many of the common species and identify them yourself. It is an enjoyable, interactive way of learning.

iNaturalist has a wealth of information on Aotearoa New Zealand Lepidoptera, with more than 175,000 observations of 1,536 species at the time of writing. If you are interested in just looking at observations in your local area, you can easily filter by region or draw a box around an area you are interested in. For example, you may be based in Christchurch and want to know what Lepidoptera live in or close to the city. A search for Lepidoptera in Canterbury District reveals more than 30,000 observations of 924 species. A more refined search of Christchurch City, drawing a box around the city from Governors Bay to Kaiapoi, reveals over 11,700 observations of 367 species. By clicking on the species tab, it can be seen that the Monarch *Danaus plexippus* is the

The Exquisite Olearia Owlet Meterana exquisita *camouflaged on a lichen-covered rock in Central Otago*

How to Use This Book

most commonly seen butterfly in Christchurch City (with more than 800 observations) and *Capua intractana* is the most commonly seen moth (with more than 260 observations).

You can search for a species on iNaturalist to see where else it has been found in Aotearoa, how often and at what time of the year. For example, at the time of writing there were 70 observations of the Exquisite Olearia Owlet *Meterana exquisita*. It can be seen on the map page that the observations are distributed from the lower North Island down to Stewart Island. There are particular concentrations near Palmerston North in the North Island and Central Otago in the South Island. By studying the species description in this book, you can see that *M. exquisita* is attracted to light and lives in habitats with small-leaved Olearia shrubs (the caterpillar's host plant). Thus, how to target the species is clear.

HOW TO USE THIS BOOK

This book covers a selection of 360 species (with photographs), representing about 20 per cent of the 1,750-odd named Lepidoptera species in Aotearoa. In terms of families, 34 of 49 are represented. Families are listed alphabetically (from Carposinidae to Zygaenidae), and in each family the species are also generally listed alphabetically. Most species are allocated a half page, with a smaller number allocated a full page. Sometimes 2–3 similar species are covered together on a page (or half page). There are also some 'genus pages', where 1–2 pages are used to cover a whole genus – with a selection of species from that genus profiled in photographs. For example, the genus *Orocrambus* (family Crambidae), contains about 51 species, and eight species are depicted in photographs across two pages in order to illustrate some of the diversity within this large genus. In cases where the genus a moth belongs to is thought to be incorrect or requires further study, this is indicated by the genus name being given in inverted commas, for example 'Megacraspedus'.

Selecting what species to include and exclude in a title covering such a large group of insects is a difficult proposition. Most of this book is devoted to what are expected to be the most commonly seen moths and butterflies. To balance the emphasis on common species, a selection of rare and particularly ecologically fascinating Lepidoptera is profiled in a special rarities and specialities section (p. 152).

The information under each species is arranged into self-explanatory sections. At the top, the common name of the species is given (where available), together with the scientific name. On the top right the wingspan is given in millimetres, as an indication of size. Below this, there are three to four sections, and a photograph or two of each species. The sections are:

- **Description** Very brief introduction to the species, including whether it is endemic, native,

The Streblus Owlet Meterana octans, *a rarely seen nocturnal moth. It is monophagous with larvae only on the Small Leaved Milk Tree* Streblus heterophyllus.

How to Use This Book

cosmopolitan and/or a migrant. Brief description of adult moth's colour, form, pattern and/or distinguishing features, as well as any stark differences between the sexes. Where useful, how to distinguish the species from similar species is at least partially explained. Sometimes a brief description of the larva is given.
- **Distribution** Location of the species across Aotearoa, including areas where it is most common. For species that also occur in other countries the wider distribution is given.
- **Habits and habitats** Description of habitats used by the species. The host plant or plants (where known) and whether the species is nocturnal (and attracted to light), or day flying. The time of year when the adults are 'on the wing' (flying around looking for mates, food or host plants). Any interesting ecological observations.
- **Remarks** Any additional interesting aspects of the species' ecology, behaviour and relationships with other species, including other Lepidoptera and humans. For rarer species, their conservation status may be given, along with a discussion on their threats or conservation requirements.

Glossary

abdomen Rearmost section of body, attached to thorax, and containing digestive, excretory and reproductive systems.
adventive Introduced and established in an area outside its usual native range; exotic.
antenna (pl **antennae**) pair of sensory organs on insect's head.
apex Tip of wing between leading edge (costa) and termen.
arboreal Living in trees for food or shelter.
basal Towards base of a structure.
binomial nomenclature Formal system used to classify organisms, whereby each species has a genus name (capitalized) and species name (not capitalized), for example *Danaus plexippus*.
biogeography Study of patterns and causes of geographical distribution of species.
camouflage Defence or tactic that organisms use to disguise their appearance, usually to blend in with their surroundings.
canopy Uppermost branches of trees in a forest, forming a distinct layer.
caterpillar Larva of a butterfly or moth.
compound eyes One of two different visual organs in insects. A compound eye does not produce a visually sharp image but gives a wide field of view and is very good at detecting movement.
cosmopolitan Found almost anywhere on Earth.
costa Leading edge, or outer edge, of moth or butterfly wing.
cremaster Terminal end of pupa, often with processes or curved setae, usually attached to silk in cocoon or on host plant.
cryptic Camouflaged by body colour, pattern and/or shape; also shy, inconspicuous behaviour.
day flying Flies during daylight hours.
diapause Period of suspended development.

■ Glossary ■

discal spots Often contrastingly coloured (pale or dark) spots in centre of wing.
diurnal Active by day.
dorsal Relating to upper surfaces of an animal's body.
ectotherm Organism that regulates its internal body temperature via external sources of heat/energy.
emergent Of an animal that is not resting and is active or visible.
endemic Native and restricted to a certain area, for example endemic to Aotearoa New Zealand.
exoskeleton Externalized skeleton that supports and protects body of an invertebrate.
extirpation Elimination/extinction of an organism from an area.
family Taxonomic rank above level of genus; usually contains a number of genera and many species.
forewings Front wings of a moth or butterfly closest to head.
genus (pl **genera**) Taxonomic rank above level of species; a genus comprises one to many related species.
gregarious Living close together in group or community.
herbaceous Relates to vascular plants lacking woody stems.
herbivorous Feeding only on plant matter.
hindwings Back wings in moth or butterfly furthest from head.
horizontal bands Markings that extend across an animal, rather than lengthways.
host plant Plant that an organism lives and feeds on.
invasive Refers to organism that causes ecological or economic harm in a new environment where it is not native.
invertebrate Refers to animal that lacks a backbone.
labial palps Important sensory organs located on each side of proboscis on head in adult.
larva (pl **larvae**) Wingless life stage of a holometabolous insect that emerges from the egg. A holometabolous insect is one which has a four-stage life-cycle, including a pupa. Also known as caterpillar.
lateral Side(s) of a structure.
Lepidoptera Order of insects comprising butterflies and moths.
lepidopterist Entomologist who specializes in studying butterflies and moths.
life cycle Series of changes that individuals of a species undergo in a single generation (that is, egg, larva, pupa and adult in Lepidoptera).
mandible Jaw-like mouthparts of many insects used for cutting and chewing.
median Located in or near middle of a structure.
migrant Refers to directional mass movement of individuals of the same species.
mimic Refers to species whose patterns and behaviour resemble that of another.
monophagous Feeding on or utilizing a single kind of food. A moth or butterfly with only one species of host plant (or a small number of related species).
native Naturally occurring or originating in a certain area.
naturalized Refers to organism introduced to a region where it is not indigenous and living wild there.
nocturnal Active by night.

Glossary

'on the wing' Refers to time of year when a moth or butterfly is actively flying as an adult.
order Taxonomic rank above family and below class.
oviparous Reproducing by laying eggs.
ovipositor Tubular structure used for laying eggs.
parasite Organism that lives on or in another for its own benefit (for example for food or shelter), harming its host.
pectinations Lateral branches on antennae of many moths. They assist in sensing chemicals in the environment and navigating towards sources of food or potential mates.
proboscis Long, mobile feeding tube extending from front of head, seen in Lepidoptera.
prothorax First segment of thorax, immediately behind head.
pupa (pl **pupae**) Life-history stage between larva and adult where reorganization into an adult occurs.
pupation Becoming a pupa.
polyphagous Of Lepidoptera with a wide range of host plants for the larvae.
pheromones Chemical signals, that is, carriers of information between individuals within a species. Pheromones induce a physiological or behavioural response in the receiving individual, and often play a crucial role in mate finding and other interactions among animals.
rare Existing in low numbers, in few locations, and/or very difficult to find places.
scree Aggregation of loose rocks on a slope that can shift when stood on.
self-introduced Refers to Lepidoptera species that made their own way to a new country, or were taken to a new country by wind, as opposed to being deliberately or accidentally moved by humans.
sexual dimorphism Refers to male and female of the same species with very different appearances.
shrubland Habitat that is dominated by shrubs rather than forest trees or grasses.
species Group of organisms that freely interbreed among themselves.
subspecies Level of taxonomic division below species.
taxon (pl **taxa**) Unit used in science for classification of organisms.
taxonomy Classification of living things.
temperate Relates to a region characterized by a mild climate.
territory Area that an individual or group occupies and defends from others.
termen Bottom edge (or outer margin) of wing that is furthest from head.
thorax Middle section of body of insect, between head and abdomen.
tornus Inner rear corner of wings, between dorsum and termen.
tubercle Raised projection.
tussockland Grassland consisting primarily of tussock.
vagrant Refers to individual that has moved well outside its usual distribution or wanders between places.
ventral Relating to surfaces on underside of an organism's body.

▪ CARPOSINIDAE ▪

Australian Guava Moth ▪ *Coscinoptycha improbana* WS 15mm

DESCRIPTION Moth with distinctive raised scales on whiteish forewings, and black or brown markings across thorax. Forewings also include dark speckling. Caterpillar pinkish and grows to more than 10mm. Can quite easily be confused with several endemic *Heterocrossa* species, but distinctive feature is 'C'-shaped dark mark about two-thirds of the way along forewing near costa. **DISTRIBUTION** Native to eastern and southern Australia. Also occurs on Norfolk Island and New Caledonia, and recorded in Aotearoa New Zealand since 1997. Established in Northland, Auckland and Waikato. **HABITS AND HABITATS** Adults on the wing year round. Guava moths lay eggs

on a large range of fruits and nuts throughout the year, including citrus, plums, peaches, pears, apples, macadamia, feijoa and guava. Moth lays its eggs on a fruit's surface, and larva bores into fruit. Larvae leave fruit and pupate when fruit has fallen to the ground. **REMARKS** The only member of its genus, *Coscinoptycha*. Feeding by larva makes the fruit inedible and causes early drop before it is fully ripe.

Heterocrossa spp. ▪ WS 14–35mm

DESCRIPTION *Heterocrossa* have distinctive tufts of raised scales on forewings. The three most commonly seen species are the Raspberry Bud Moth *H. rubophaga*, *H. exochana* and the Lichen Snoutlet *H. eriphylla*. Raspberry Bud Moths variable, some being light in colour, others darkish-brown or grey. *H. exochana* whitish-brown with elongated forewings. Lichen Snoutlets whitish-green with black, brown and green markings. **DISTRIBUTION** Both the Raspberry Bud Moth and *H. exochana* widespread but infrequently encountered. The Lichen Snoutlet is widespread in the North Island and upper South Island. **HABITS AND HABITATS** All species attracted to light. Raspberry Bud Moth larvae feed on raspberry and blackberry. The Lichen Snoutlet lives in native forest and larvae feed on wood of beech trees *Fuscospora* and wineberry *Aristotelia serrata*. Larvae of *H. exochana* feed on fruits of *Muehlenbeckia* species. **REMARKS** The endemic genus *Heterocrossa* contains about 17 species that can be difficult to identify.

Raspberry Bud Moth (left), Heterocrossa exochana *(centre)*, *Lichen Snoutlet (right)*.

CHOREUTIDAE/COSMOPTERIGDAE

Small Thistle Moth ■ *Tebenna micalis* WS 13mm (e)

DESCRIPTION Very distinctive small moth with metallic markings and broad, rounded wings. **DISTRIBUTION** Native, self-introduced moth, found worldwide (cosmopolitan). Present throughout Aotearoa New Zealand but rare in Southland and cold parts of the South Island. **HABITS AND HABITATS** Larvae usually recorded on plants in daisy family (Asteraceae). Known larval food plants in New Zealand include native woollyheads *Craspedia*, Capeweed *Arctotheca calendula*, Scotch Thistle *Cirsium vulgare*, Horseweed *Erigeron canadensis*, the thistle *Onopordum acanthium* and the Golden Everlasting *Xerochrysum bracteatum*. Larvae skeletonize the leaves of their host and pupation occurs in a silk cocoon under a leaf. Adults come to light and can be found by day on flowers of larval food plant. **REMARKS** There are more than 20 other species in the Choreutidae family (metalmark moths) in New Zealand. They are all endemic and belong to the genus *Asterivora*, which is especially diverse in alpine areas.

Pyroderces apparitella ■ WS 8mm (e)

DESCRIPTION Unusual moth with bright red eyes and bright silvery-white curved lines across elongated forewings that are bordered in black. Endemic moth in the Cosmopterigidae family. **DISTRIBUTION** Fairly common throughout the North Island. Also present in the Nelson region of the upper South Island. **HABITS AND HABITATS** Preferred habitat is native forests and residential gardens. Larvae feed in thin, dead branches of many plants, including tutu *Coriaria* spp. and Supplejack *Ripogonum scandens*. Adults on the wing in October–February and attracted to light. When resting, wings curve upwards away from substrate.

CRAMBIDAE

Cotton Web Spinner ■ *Achyra affinitalis* WS 20mm

DESCRIPTION Adult moth has fawn to dark brown forewings, and pale brown hindwings with satin sheen. Forewings sometimes suffused with grey and with or without two irregular blackish cross-lines. **DISTRIBUTION** Self-introduced migrant. Native to Australia. Very common species in eastern Australia and probably the most widespread and common crambid there. First recorded in Aotearoa New Zealand in 1973 and now widespread and established in the North and South Islands. **HABITS AND HABITATS** Larvae feed on the Common Sunflower *Helianthus annuus*, Lucerne *Medicago sativa*, Linseed *Linum usatissimum*, *Sorghum bicolor* and saltbushes *Atriplex* spp. Can be seen flying by day in grassland and shrubland, but also comes to light at night.

Antiscopa epicomia ■ WS 18–19mm

DESCRIPTION Endemic genus. Contains three species. *A. epicomia* is depicted. Forewings pale light ochreous-grey with basal third being reddish-brown. Hindwings pale whitish-grey. Species varies in size and in colour and intensity of markings. Very similar in appearance to *A. acompa*, but *A. acompa* has a thicker antemedian forewing line. Third species in genus, *A. elaphra*, is a smaller moth (wingspan 12–14mm), light fawn in colour with scattered black dots; it is common in grassland and shrubland. **DISTRIBUTION** *A. epicomia* found throughout Aotearoa New Zealand, including on Auckland Island, Campbell Island and the Kermadec Islands. **HABITS AND HABITATS** Inhabits native forest, preferring damp, shaded areas. Adults variable in size and colouration; on the wing all year but most often in October–April. Attracted to light. Life history unknown.

CRAMBIDAE

Culladia cuneiferellus ■ WS 10mm

DESCRIPTION Pale brown moth with two dark chevrons on forewings. **DISTRIBUTION** Self-introduced migrant since 1999. Now widespread across the North Island but most common in the upper North Island and especially common in coastal areas. Also recorded in Nelson district at top of the South Island. Additionally found in Australia (Queensland, New South Wales and Tasmania), New Caledonia, Norfolk Island, the New Hebrides and the Loyalty Islands. **HABITS AND HABITATS** Larvae feed on various grasses on lawns and pastures.

Clematis Triangle ■ *Deana hybreasalis* WS 25–30mm (e)

DESCRIPTION The sexes differ in this beautiful, charismatic species. Males usually pinkish-brown or purplish-brown and females pale orange to golden-brown. Both sexes have bright white markings along forewing edge, which, along with wing shape, help distinguish this species. The only species in the endemic moth genus *Deana*. **DISTRIBUTION** Widespread and fairly common across the North and South Islands. **HABITS AND HABITATS** Inhabits native forest and shrubland. Larvae known to feed on leaves and flowers of *Clematis* vine species. Flies at night and comes to light. On the wing throughout the year, but less commonly encountered in winter.

■ CRAMBIDAE ■

Arrowhead ■ *Diasemia grammalis* WS 13–15mm ⓔ

DESCRIPTION Bright little moth endemic to Aotearoa New Zealand. When forewings are rested over hindwings, forms a distinctive arrowhead shape. Brown and black in colour, adorned with white bands or patches. May have bright orange or yellow streaks on forewings. **DISTRIBUTION** Widespread and reasonably common across the North and South Islands. **HABITS AND HABITATS** Active day-flying moth in warm, sunny weather. Often sighted flying low over short-stature vegetation. Larvae feed on native grasses and herbaceous plants. Seen flying by day in grassland and shrubland, around river terraces, lake edges, wetlands and mires.

Eastern Black Tabby ■ *Diplopseustis perieresalis* WS 12–20mm

DESCRIPTION Moth with ability to tolerate a wide range of climates, from subantarctic islands to the tropics. **DISTRIBUTION** Widespread and reasonably common across the North and South Islands. Also present on Rakiura/Stewart Island, the Chatham Islands and the Antipodes Islands. Globally widespread. Native to parts of Asia, Australia and Aotearoa New Zealand. Has also been introduced to Europe, where it is now widespread.

HABITS AND HABITATS Present in native forest, shrubland and sedgeland. Adults can be found throughout the year and come to light. Can also be disturbed from vegetation by day. Larvae thought to feed on the endemic sedge Makura *Carex secta* and other large sedges. Comes to light. **REMARKS** Can be abundant but is not often seen. May be found at night with a torch, searching areas with sedges.

CRAMBIDAE

Eudonia spp. ■ WS 13–27mm

DESCRIPTION Globally widespread genus in subfamily Scopariinae. Roughly 250 species are currently placed in the genus and new species are still being described. Many species rather inconspicuous in greys or browns, though some are more striking, such as the aptly named Bold Scoparia *E. aspidota*. At least 57 species endemic to Aotearoa New Zealand. To display some of the variation within the genus and outline the most common species likely to be encountered across the country, 12 species are profiled here: the Bold Scoparia *E. aspidota*, *E. submarginalis*, the Stone Moth *E. cataxesta*, *E. philerga*, *E. melanaegis*, *E.*

Bold Scoparia Eudonia aspidota

Eudonia submarginalis

Stone Moth Eudonia cataxesta

Eudonia philerga

Eudonia melanaegis

Eudonia steropaea

CRAMBIDAE

steropaea, *E. sabulosella*, *E. octophora*, *E. leptalea*, *E. feredayi*, the Shining Scoparia *E. diphtheralis* and *E. trivirgata*. **DISTRIBUTION** Above 12 species all widespread across New Zealand in both the North and South Islands. Some also occur on Rakiura/Stewart Island, the Chatham Islands and the Auckland Islands. **HABITS AND HABITATS** As far as is known, larvae of most *Eudonia* species feed on mosses. Some also eat lichen. In a few cases, other food plants have been recorded or suspected, including grasses, herbs and bryophytes. Adults of some *Eudonia* species have been seen visiting the flowers of native shrubs such as mānuka *Leptospermum* spp., *Olearia virgata*, *Helichrysum intermedium* and *Dracophyllum acerosum* – probably feeding from them and pollinating them. Most species nocturnal and come to light. Many are also active fliers by day.

Eudonia sabulosella

Eudonia octophora

Eudonia leptalea

Eudonia feredayi

Shining Scoparia Eudonia diphtheralis

Eudonia trivirgata

CRAMBIDAE

Gadira acerella ■ WS 16–18mm

DESCRIPTION Distinctively patterned and coloured, a characteristic that is thought to camouflage the moth against rocks and lichens. Rests with wings together over body in a steep 'V' shape. Unlikely to be confused with any other Lepidoptera species in Aotearoa New Zealand. **DISTRIBUTION** Endemic, widespread and fairly common across the North and South Islands. **HABITS AND HABITATS** Inhabits native forest and shrubland from sea level to subalpine altitudes. Larvae thought to feed on lichen or moss. Adults on the wing in October–March. Active at night and attracted to light.

Swan Plant Flower Moth ■ *Chabulina onychinalis* WS 15mm

DESCRIPTION Moth with distinctive striking grey, brown and white pattern. **DISTRIBUTION** Introduced to Aotearoa New Zealand. Widespread across the North Island (most common in Auckland and Waikato regions). Rare in the South Island. Widely distributed in Africa, India, South-east Asia and Australia. Also recorded in North America (California) since 2000. **HABITS AND HABITATS** Larvae recorded feeding on the Swan Plant *Gomphocarpus fruticosus*, and Oleander *Nerium oleander*, and probably also feed on other plants. Attracted to light.

CRAMBIDAE

Glaucocharis spp. ■ WS 11–17mm

DESCRIPTION Widespread moth genus, with more than 150 species occurring in Australia, India, South-east Asia, America and Africa. At least 18 are endemic to Aotearoa New Zealand. Characteristic unusual triangle shape and an array of colours and patterns. Four species are profiled here, displaying some of the variation within the genus: the Yellow Silverling G. *auriscriptella*, G. *elaina*, G. *lepidella* and G. *pyrsophanes*.
DISTRIBUTION All four species widespread across New Zealand. **HABITS AND HABITATS** *Glaucocharis* moths occur in a wide variety of habitats, including wetlands, coastal areas, shrubland and forest. Their larvae feed on mosses and liverworts. Adults of G. *pyrsophanes* are thought to pollinate mānuka, *Leptospermum* spp., and *Helichrysum selago*. Many *Glaucocharis* moths come to light at night. Some also actively fly by day or can be seen by disturbing vegetation, causing the moths to take flight. Adults actively on the wing in warmer months of the year in September–April.

Yellow Silverling Glaucocharis auriscriptella

Glaucocharis elaina

Glaucocharis pyrsophanes

Glaucocharis lepidella

CRAMBIDAE

Pond Moth ■ *Hygraula nitens* WS 14–16mm

DESCRIPTION Pond Moths have brown forewings with white markings and patchy, buff-coloured hindwings. Larvae notable for being the only native aquatic caterpillar in Aotearoa New Zealand. They are among the few freshwater invertebrates in the country that feed on aquatic plants. Pale yellowish or greenish larvae easily distinguished from terrestrial caterpillars by clusters of tentacle-like gills along body. **DISTRIBUTION** Native and widespread in both New Zealand and Australia. **HABITS AND HABITATS** Larvae live underwater, and thus have gills. They are delicate and build leaf housing for shelter and protection. Larvae feed on various native and introduced water plants, including the Curly-leaf Pondweed *Potamogeton crispus*, eel grasses *Zostera* spp., Hydrilla *Hydrilla verticillata*, Curly Waterweed *Lagarosiphon major*, Hornwort *Ceratophyllum demersum*, Canadian Pond Weed *Elodea canadensis*, *Myriophyllum propinquum* and Red Pondweed *Potamogeton cheesemanii*. Adult moths frequently come to light close to waterways.

Poroporo Fruit Borer ■ *Leucinodes cordalis* WS 20–34mm

DESCRIPTION Pretty moth with off-white forewings and pattern of pale brown or orange-brown blotches. **DISTRIBUTION** Native species shared with Australia and Indonesia. Widespread throughout the North Island, less so in the South Island, and generally coastal. Common where its host plants occur. **HABITS AND HABITATS** Frequents forest edges, parks and gardens, coastal dunes and weedy places. In Aotearoa New Zealand larvae feed on the native Poroporo *Solanum aviculare* and Glossy Nightshade *S. americanum*. In addition, larvae utilize many introduced plants and vegetables such as eggplant, pepino, tomato, potato and chili pepper. They bore into the fruits of their host plant and feed on the flesh and seeds. Adult moths fly at night and come to light. On the wing in October–May.

CRAMBIDAE

Rusty Dotted Triangle ■ *Mnesictena flavidalis* WS 17–24mm e

DESCRIPTION Small, brightly coloured endemic moth. Rather variable in colour but pattern is consistent. Some individuals bright orange with yellow hindwings, others dusky-brown. **DISTRIBUTION** Widespread in both the North and South Islands. Abundant in Otago and Canterbury. **HABITS AND HABITATS** Occurs in a variety of habitats, including native forests, shrubland, gardens and coastal areas. Host plants include Pōhuehue

Muehlenbeckia complexa. Comes to light and can also be seen flying by day. Mostly on the wing in October–April.

Golden-brown Fern Moth ■ *Musotima nitidalis* WS 15–25mm

DESCRIPTION Native to Australia and Aotearoa New Zealand. Adult moths brown or orange-brown with various bright white markings outlined in black or dark brown on each forewing. **DISTRIBUTION** Known from New Zealand and most of Australia. Widespread in New Zealand. Present on the North, South, Rakiura/Stewart and Chatham Islands; also on the subantarctic Antipodes Island. **HABITS AND HABITATS** Larvae feed on undersides of leaves of various ferns, including Bracken *Pteridium esculentum* and Mātā *Histiopteris incisa*. Caterpillars live in a sparse web. Pupation takes place in a folded leaf of the host plant held by silk. **REMARKS** In 2009 the moth was found in Dorset, England, and since then the species has become established in southern England. It is suspected to have been introduced in imported ferns.

CRAMBIDAE

Orocrambus spp. ■ WS c. 15–30mm

DESCRIPTION *Orocrambus* is a genus of grass moths with more than 50 species, all endemic to Aotearoa New Zealand. Many have long, elongated forewings, but they vary widely in shape, size and patterning. Some look quite similar to each other, thus clear photographs from several angles and expert opinion may be needed to identify species. To display some of the variation within the genus and outline the most common species likely to be encountered, eight species are profiled here: *O. vittellus*, the Common Grass Moth *O. flexuosellus*, *O. vulgaris*, *O. ramosellus*, the Mire Grass Moth *O. aethonellus*, *O. corruptus*. *O. cyclopicus* and *O. angustipennis*. Wingspan of most species is 15–30mm, but *O. angustipennis* is larger (34–50mm), and *O. aethonellus* is small (13–21mm). **DISTRIBUTION** *O. flexuosellus*, *O. vittellus*, *O. ramosellus* and *O. angustipennis* all widespread across New Zealand in both the North and South Islands. Some of the species also occur on Rakiura/Stewart Island, the Chatham Islands and even the subantarctic islands. *O. vulgaris* and *O. cyclopicus* occur in the lower-central North Island and across the South Island. *O. corruptus* is found in dry parts of Canterbury and Otago. *O. aethonellus* is widespread in damp habitats of the South Island. **HABITS AND HABITATS** *Orocrambus* moths are common in lowland grassland, town gardens, coastal areas, wetlands, shrubland, subalpine

Orocrambus vittellus

Common Grass Moth Orocrambus flexuosellus

Orocrambus vulgaris

Orocrambus ramosellus

CRAMBIDAE

Mire Grass Moth *Orocrambus aethonellus*

Orocrambus corruptus

Orocrambus cyclopicus

Orocrambus angustipennis

tussock grassland, herb fields, and forest edges and clearings around the country. Many species are also present in the alpine zone, and some are restricted to it. Species such as *O. corruptus* appear to thrive in very dry, poorly drained areas, including lawns and old pastures, whereas those like the Mire Grass Moth thrive in wetlands and bogs. This illustrates the wide range of habitats exploited by the genus. Larval host plants vary. Larvae generally oligophagous on grasses, sedges or rushes, and some species also eat mosses. The large *O. angustipennis* larvae feed on Toetoe *Austroderia toetoe* and Common Pampas Grass *Cortaderia selloana*. Many adults visit flowers of native plants and may pollinate them. Many *Orocrambus* moths will come to light at night, but many also actively fly by day, such as *O. flexuosellus* and *O. vittellus* in grassland. Some can be seen by disturbing vegetation by day, causing the moths to take flight. Adults actively on the wing primarily in warmer months in October–April.

CRAMBIDAE

Scoparia spp. ■ WS 15–31mm

DESCRIPTION Large and globally widespread genus in the subfamily Scopariinae. More than 230 species in genus, which occurs on every continent except Antarctica. At least 58 species are endemic to Aotearoa New Zealand, although some of these may belong in the genus *Eudonia* and more taxonomic work is required on the country's Scopariinae. Some species rather inconspicuous in greys or browns. Others more striking, such as the Boot Scoparia *Scoparia ustimacula*, or have strongly contrasting black and white markings, like *S. rotuella*. To display some of the variation within the genus and outline the most common species likely to be encountered in New Zealand, eight species are profiled here: *S. subita*, *S. exilis*, *S. petrina*, the Boot Scoparia *S. ustimacula*, *S. rotuella*, *S. halopis* and *S. minusculalis*. In addition, a beautiful golden *Scoparia* species that is currently undescribed is depicted. This species is referred to as *Scoparia* s.l. sp. A in the New Zealand Arthropod Collection (NZAC). **DISTRIBUTION** *S. halopis*, *S. minusculalis*, the Boot Scoparia, *S. rotuella* and *S. petrina* all widespread in the South Island. In the North Island, *S. halopis*, *S. minusculalis* and the Boot Scoparia are widespread, *S. rotuella* widespread as far north as the southern Auckland district, and *S. petrina* to the central North Island only. Some species also occur on Rakiura/Stewart Island, the Chatham Islands and even the subantarctic

Scoparia subita

Scoparia s.l. sp. A

Scoparia exilis

Scoparia petrina

CRAMBIDAE

Boot Scoparia Scoparia ustimacula

Scoparia rotuella

Scoparia halopis

Scoparia minusculalis

islands. The beautiful day-flying *S. subita* is only present in alpine areas of the southern South Island. Finally, *S. exilis* is widespread in the South Island only. **HABITS AND HABITATS** Larval host plants vary and are often unknown. However, they include mosses, ferns, tussocks and flowering plants. Boot Scoparia have larvae on *Hydrocotyle*, and *S. rotuella* on willowherbs *Epilobium*. Adults of some *Scoparia* species, such as *S. rotuella*, have been seen visiting the flowers of native shrubs, and possibly pollinating them, such as mānuka, *Leptospermum* spp., *Helichrysum* spp. and *Veronica salicifolia*. Many *Scoparia* species are nocturnal and come to light. Some of the nocturnal ones can be seen by disturbing vegetation by day, causing the moths to take flight. Some species, especially in the alpine zone, such as *S. subita*, are also active day fliers.

CRAMBIDAE

Proternia philocapna ■ WS 21–26mm

DESCRIPTION Endemic moth and the only member of its genus. Forewings dark grey with pale markings and darker lines. Males have a characteristic bend or kink in their antennae, but females lack this modification. **DISTRIBUTION** Widespread in the North Island. **HABITS AND HABITATS** On the wing November–March. Attracted to light. Larvae have been found feeding on leaf litter from silk runways under logs in native forest.

Kowhai Moth ■ *Uresiphita maorialis* WS 22–27mm

DESCRIPTION Eye-catching moth with bright orange and black hindwings. Forewings vary from reddish-brown or olive-green to dark grey. Hindwings also vary, being bright orange or yellow, but also occurring in olive-green, brown or grey. Black band at outer margin of hindwings. Larvae conspicuous, being bright green spotted with black and white. A very similar species, *U. ornithopteralis*, can easily be confused with this one; it is native to Australia and is thought to occur here entirely (or mostly) as a migrant. *U. ornithopteralis* can be distinguished from the Kowhai Moth based on having a definite black border on underside of hindwings, which is reduced in the latter species. **DISTRIBUTION** Reasonably common and widespread in the North, South and Rakiura/Stewart Islands. Overlaps in distribution with *U. ornithopteralis* in Auckland, Northland and potentially elsewhere. **HABITS AND HABITATS** Larvae feed on endemic kowhai, *Sophora* spp., as well as the introduced Tree Lupin *Lupinus arboreus* and Gorse *Ulex europaeus*. They occur in forest, shrubland and coastal dunes. The moth also occurs in town and city gardens

wherever kowhai has been planted, or occurs naturally. Larvae well known as a defoliator of kowhai trees, but they do not harm or kill the plants and the insect is a lovely moth to see in the garden. Sometimes flies by day with a spinning flight. Also flies at night and comes to light. **REMARKS** *U. ornithopteralis* may be established at low levels in the northern North Island, and there are even presumed hybrids between the two species.

DEPRESSARIIDAE

Poison Hemlock Moth ■ *Agonopterix alstromeriana* WS 17–19mm

DESCRIPTION Nocturnal species accidentally introduced to Aotearoa New Zealand. Adult moths pale brown with lighter coloured area near bases of wings and a few darker spots near edges of wings. Distinguishing brown or reddish spot near centre of each wing. **DISTRIBUTION** Widespread species native to Europe. Accidentally introduced to a number of areas, including the United

States, southern Canada, northern Europe and, more recently, New Zealand and Australia. Has been breeding in New Zealand since 1986. **HABITS AND HABITATS** Used in biological control of the lethally toxic Poison Hemlock *Conium maculatum*, which is eaten by its larvae. Caterpillars spin leaves of host plant into a protective tube, in which they live. Found wherever its host plant occurs. Thus, the moth is most commonly seen in open fields and roadsides in suburban and rural locations. Adults attracted to light.

Gorse Soft Shoot Moth ■ *Agonopterix umbellana* WS 21mm

DESCRIPTION Native to Europe but was introduced to Hawaii in 1988 and Aotearoa New Zealand in 1990 to control the invasive Gorse *Ulex europaeus*. Forewings whitish-

ochreous with dark brown streaks. Hindwings pale whitish-grey. **DISTRIBUTION** Widespread in New Zealand following the distribution of Gorse. Most common in the lower North Island and eastern side of the South Island. **HABITS AND HABITATS** Adults on the wing in August–April and attracted to light. Hibernates in winter and can reappear in early spring. **REMARKS** Larvae feed on Gorse within silken tubes.

▪ DEPRESSARIIDAE ▪

Eutorna symmorpha ▪ WS 12–14mm e

DESCRIPTION Endemic moth with yellow-brown or reddish-brown forewings, sometimes streaked with whitish-brown between reddish or brownish veins. Conspicuous small black mark or spot on forewings. Hindwings light grey. **DISTRIBUTION** Widespread on the North, South and Chatham Islands. **HABITS AND HABITATS** Mostly on the wing in October–April. Nocturnal and attracted to light. Life history unknown. **REMARKS** Depressariidae is a family of moths comprising about 2,300 species worldwide.

Tarata Flat Moth ▪ *Nymphostola galactina* WS 23–26mm e

DESCRIPTION Gorgeous white moth of forested habitats, delicately tinged with green. Head, palpi, antennae, thorax, abdomen and legs all snow-white, tinged with pale green. Forewings broad, and costa strongly arched. Hindwings and cilia white. A very distinct species. *Nymphostola* is a monotypic endemic moth genus. **DISTRIBUTION** Uncommon to fairly common on the North and South Islands. Rare in the upper North Island. **HABITS AND HABITATS** Larval food is the leaves of *Lophomyrtus bullata*, Kōhūhū *Pittosporum tenuifolium* and Akapuka *Griselinia lucida*. Larvae live on uppersides of leaves under a fine silk web. Adults attracted to light, mostly in November–February.

DEPRESSARIIDAE

Phaeosaces apocrypta & *P. coarctatella* ■ WS 22–30mm

Phaeosaces apocrypta

DESCRIPTION Both of these endemic species are variable in colour. *P. apocrypta* varies from light whitish-brown to dark grey or brown. *P. coarctatella* variable in appearance, with some individuals having a greenish shade to forewings and others being reddish-brown. Extent of black shading on forewings also varies. Both species have two widely spaced dark spots in central area of each forewing, but these are not always clearly visible. **DISTRIBUTION** *P. apocrypta* widespread from the central North Island southwards; most common in the South Island, particularly Otago and Canterbury. *P. coarctatella* widespread in the North and South Islands. **HABITS AND HABITATS** Larvae of both species hide in hollow twigs, under bark or in old wood-borer tunnels by day. They are nocturnal and emerge at night to feed on lichens growing on tree bark during the evening. Larvae pupate in their shelter. Both species live in native forests and shrubland, but may also turn up in domestic gardens. *P. apocrypta* mostly on the wing in November–March and *P. coarctatella* in September–January. Both species nocturnal as adult moths and attracted to light. **REMARKS** The phallus (penis) of male *P. coarctatella* is very long, taking up most of the length of the abdomen, and having more than 30 spirals.

Phaeosaces coarctatella

▪ ELACHISTIDAE ▪

Typical grass miner moths ▪ *Elachista* spp. WS 8–15mm

DESCRIPTION *Elachista* are small to very small moths (wingspans usually around 1cm). Their wings appear feather-like, and hindwings are significantly reduced. Pictured here is an unknown species (could not be easily identified beyond genus level) and *E. thallophora*. The latter has pearly-white forewings with distinctive brown longitudinal streaks. **DISTRIBUTION** Widespread on the North, South and Chatham Islands. Essentially found worldwide, except in very cold places and on some oceanic islands. **HABITS AND HABITATS** Larvae typically leaf or stem miners on grasses, rushes and sedges. *E. thallophora* common in tussock grassland in subalpine and alpine zones up to at least 1,600m.

Elachista *sp.*

Elachista thallophora

Microcolona limodes ▪ WS 7–8mm 🟢

DESCRIPTION Endemic to Aotearoa New Zealand. Forewings very narrow and brown and white, sprinkled with black. Three or more raised tufts of black scales on either side of moth, giving it a 'bumpy' appearance. Hindwings whitish-grey. **DISTRIBUTION** Widespread and can be common but easily overlooked on both the North and South Islands. **HABITS AND HABITATS** Larvae eat seeds of the Red Mapou *Myrsine australis* and Toro *M. salicina*.

EREBIDAE

Northern Wattle Moth ■ *Dasypodia cymatodes* WS 80mm

DESCRIPTION Large and impressive moth with a very large wingspan. A similar species, the Southern Wattle Moth *D. selenophora*, also occurs widely in Aotearoa New Zealand but is rarer. The Southern Wattle Moth has bigger eye-spots on the wings and large zigzag lines on the inside of these. Spikes on zigzags bigger than on the Northern Wattle Moth, and there are three distinct points. In addition, the Southern Wattle Moth has bright yellow scaling on the front of the thorax and underside of the body and wings. **DISTRIBUTION** Immigrant and self-introduced resident native to Australia. Widespread on the North Island and upper half of the South Island. Most common in upper North Island. **HABITS AND HABITATS** Occupies parks and gardens, weedy areas and roadsides. Larvae feed on wattles *Acacia* spp. Nocturnal and comes to light. On the wing mostly in November–April. Moths survive over winter as adults and sometimes enter houses in autumn seeking overwintering sites.

Pantydia sparsa ■ WS 40mm

DESCRIPTION Easily identified by wing shape and black 'collar' on thorax behind eyes. First recorded in Aotearoa New Zealand in 2004, when it probably arrived as an immigrant

from Australia. Now well established in northern New Zealand and spreading southwards, reaching the South Island in recent years. **DISTRIBUTION** Widespread on the North Island and recently recorded in the upper South Island. Most common in upper North Island. **HABITS AND HABITATS** Occupies parks and gardens, weedy areas, coastal dunes and forest edges. Larvae are polyphagous and thus feed on various shrubs and herbaceous plants. Possibly favours legumes (Fabaceae) and has been reared in captivity in New Zealand on *Lotus pedunculatus*. Nocturnal and comes to light. On the wing throughout the year.

▪ EREBIDAE ▪

Magpie Moth/Mokarakara ▪ *Nyctemera annulata* WS 35–45mm (e)

DESCRIPTION Endemic to Aotearoa New Zealand. Adult moth has black wings with white markings. Thorax and abdomen black with bands of yellow-orange. Forewings have two white spots that are more elongated, while hindwing only has single spot near to centre. Very similar to the self-introduced resident from Australia *N. amicus*. However, *N. amicus* has narrow yellow fringes around wings and orange spot on front of head. The Magpie Moth's larvae are about 35–38mm long when fully grown, and are mostly black

with lines of red, as well as blue spots and tufts of hair on each segment. **DISTRIBUTION** Found all over New Zealand, in both the North and South Islands, as well as on many of the smaller outlying islands (including Rakiura/Stewart Island, Chatham Islands and the Antipodes Islands). **HABITS AND HABITATS** Diurnal moth. Because it flies by day and has a striking appearance, it is often mistaken for a butterfly. Adults occur in coastal areas, grassland, weedy areas, roadsides and gardens. Larvae feed on *Senecio* (a genus of flowering plants), including the introduced weed Ragwort *Jacobaea vulgaris* and Common Groundsel *Senecio vulgaris*. New Zealand has 19 native *Senecio* species and 14 exotic ones, many of

which provide food sources for Magpie Moths. Moths most active in September–June in the mornings and evenings, hovering over their host plants. **REMARKS** *N. amicus* is a relatively recent arrival in New Zealand and is able to hybridize with the endemic Magpie Moth.

Larva

Pinion-streaked Snout ▪ *Schrankia costaestrigalis* WS 16–22mm

DESCRIPTION Small, with dark patch in centre of forewings. This species closely resembles moths of the Crambidae family.
DISTRIBUTION Widespread on the North, South and Chatham Islands. Also found in many countries in Europe, the Canaries, Madeira, Syria and Armenia.
HABITS AND HABITATS Occupies forests and shrubland. Larva unknown in Aotearoa New Zealand, but overseas feeds mainly on flowers of herbaceous plants. Nocturnal and comes to light.

EREBIDAE

Slender Owlet Moth ■ *Rhapsa scotosialis* WS 35mm ⓔ

DESCRIPTION Well-known and often seen endemic moth. Both sexes vary in colour, being light to dark brown, reddish, purplish or almost black. Wing shape and elongated palps distinctive. **DISTRIBUTION** Widespread on the North, South and Rakiura/Stewart Islands. One of the most common moths found in forests and shrubland. **HABITS AND HABITATS** Occupies forests and shrubland, also venturing into towns and vegetated urban environments. Larvae have been seen feeding on fallen leaves of various native trees and shrubs, and on mosses. Adults on the wing throughout the year. Nocturnal and will come to light. Seen at night feeding on White Climbing Rātā *Metrosideros diffusa* blossoms.

Cinnabar Moth ■ *Tyria jacobaeae* WS 32–42mm

DESCRIPTION Cinnabar Moths are beautiful day-flying insects with distinctive pinkish-red and black wings. The moth is named after the red mineral, cinnabar, due to the red patches adorning its jet-black wings. Larvae pale yellow initially, but later develop to jet-black with orange/yellow stripes. They are gregarious, can grow up to 30mm and are ravenous eaters. **DISTRIBUTION** Common across much of Aotearoa New Zealand, although rare in southern half of the South Island. Native to Europe and western and central Asia. Introduced into New Zealand, Australia and North America to control the Common Ragwort *Jacobaea vulgaris*, on which the larvae feed. **HABITS AND HABITATS** Larvae feed on ragwort. Cinnabar Moths fly by day but are also often attracted to light at night.

GELECHIIDAE

Anisoplaca achyrota ■ WS 17–1mm (e)

DESCRIPTION Endemic moth species. Forewings light brownish or greyish, interspersed with whitish and blackish scales. Two small black discal spots, surrounded by whitish rings.

DISTRIBUTION Widespread and has been seen in both the North and South Islands. **HABITS AND HABITATS** Inhabits native forests. Larvae feed on the green seeds of *Hoheria* species. Readily comes to light. Adults commonly on the wing in December–February.

Stone Jumper ■ *Kiwaia lithodes* WS 16mm (e)

DESCRIPTION Wonderful blueish-grey moth that appears very camouflaged on a mountainous riverbed. Endemic to Aotearoa New Zealand. Forewings grey, subtly sprinkled with blue and whitish scales. Two discal spots and one on the fold are faintly darker. Hindwings whitish-grey. **DISTRIBUTION** Widespread in the South Island in both coastal and inland locations. Can occur above tree line in the low alpine zone as well as in low-altitude situations. Also present on the Wellington coastline in lower North Island. **HABITS AND HABITATS** Larvae thought to feed on the native groundcover plant Scabweed *Raoulia australis*. The species' blue-grey-white colouring blends in well with this plant, which often grows over rocks, with which the moth is also well camouflaged.

REMARKS The genus *Kiwaia* in the family Gelechiidae contains 25 currently recognized species in New Zealand, as well as 14 species from the eastern Palaearctic (Russia, India, China and Nepal). These moths are generally small and have an interesting array of colours and patterns.

GELECHIIDAE

'*Megacraspedus*' *calamogonus* ■ WS 10–16mm (e)

DESCRIPTION Endemic moth with whitish-brown forewings, darker brown streaks and occasional dark spots. Very well camouflaged in tussock grassland. **DISTRIBUTION** Widespread in the central-lower North Island, the South Island and Rakiura/Stewart Island. Can be locally abundant. **HABITS AND HABITATS** Larvae feed on flowers and seeds of grasses. Hosts include Richard's Toetoe *Austroderia richardii*, and snow tussocks in the genus *Chionochloa*. Attracted to light and can also be easily found by day in tussock grassland. Often overlooked due to its small size and because it is well camouflaged among tussock grassland.

Tomato Stemborer ■ *Symmetrischema tangolias* WS 20mm

DESCRIPTION The Tomato Stemborer is native to South America but has spread worldwide. Adult moth fawn with large dark mark on costa of each forewing. Caterpillar fawn with stripes and darker head. **DISTRIBUTION** Widespread on the North and South Islands. Also found in the United States, South America (Peru, Bolivia, Colombia, Ecuador and Chile), Australia and Indonesia. **HABITS AND HABITATS** Larvae feed on potatoes *Solanum tuberosum*, tomatoes *S. lycopersicum* and other species of *Solanum*. Larvae feed on tubers as well as stems and leaves of plant. Pupation takes place among the debris of the host plant.

◾ GEOMETRIDAE ◾

Golden Grass Carpet ◾ *Anachloris subochraria* WS 30mm

DESCRIPTION Beautiful yellow moth native to Australia and Aotearoa New Zealand. Forewings usually light brown, yellow or tan. Each forewing's inner margin has dark quadrant patch. Dark band running across forewing and small dark discal dot. **DISTRIBUTION** Throughout New Zealand and parts of Australia. **HABITS AND HABITATS** In New Zealand larvae known to feed on the Common Ragwort *Jacobaea vulgaris*. The moths have also been reared on willowherb *Epilobium* spp., in captivity in New Zealand, and the larvae are well camouflaged among the seedpods. Day flying, but also comes to light, and found on the wing in November–April.

Asaphodes abrogata & *A. aegrota* ◾ WS 22–26mm

DESCRIPTION Two subtle-yellow endemic moths. In *A. abrogata* both the forewings and hindwings are light yellow through to dark yellow. Hindwings often paler; sometimes a transverse line near termen of forewings. *A. aegrota* variable in intensity of markings on forewings. The major difference is that *A. abrogata* always has dark shading along the forewing termen, which is lacking in *A. aegrota*. *A. abrogata* also usually much brighter orange-yellow. **DISTRIBUTION** *A. abrogata* found in the South Island. Most widespread in eastern Southland, Canterbury and Otago. *A. aegrota* occurs in the North, South and Rakiura/Stewart Islands. **HABITS AND HABITATS** *A. abrogata* on the wing in February–April and attracted to light. Inhabits open country from the coast to 1,600m above sea level. Larvae of *A. aegrota* feed on native herbs. Adults inhabit open spaces in lowland native forest, shrubland and tussock grassland. Adults on the wing in November–March. **REMARKS** Planting the endemic species *Plantago raoulii* is recommended to attract them.

Asaphodes abrogata

Asaphodes aegrota

GEOMETRIDAE

Dotted Green Carpet ■ *Asaphodes beata* WS 22–26mm

DESCRIPTION Attractive endemic moth with variable white markings on forewings. Similar to the rarer Bright Green Carpet *A. philpotti* but this lacks discal spots. **DISTRIBUTION** Throughout Aotearoa New Zealand. Common in coastal areas. Rare in areas that lack native forest. **HABITS AND HABITATS** Inhabits native forest and shrubland. Larvae have been found on native herbs in the genera *Epilobium*, *Cardamine* and *Stellaria*, and also on introduced Watercress *Nasturtium officinale*. Adults attracted to the flowers of the White Rātā *Metrosideros diffusa*. They are on the wing year round, but less commonly in winter. **REMARKS** The endemic genus *Asaphodes* contains nearly 50 species, many of which are rare alpine moths.

Elegant Carpet ■ *Asaphodes chlamydota* WS 22–28mm

DESCRIPTION Distinctive endemic moth unlikely to be confused with any other species. Forewings light yellow-brown, with two broad, dark, purplish-brown bands. Outer band strongly curved. Hindwings light orange strongly suffused with blackish in most North Island populations, whereas in South Island populations they are a light orange with little pattern. **DISTRIBUTION** Found throughout most of the country, from Northland to Southland. Particularly common in northern Central Otago and South Canterbury. **HABITS AND HABITATS** Predominantly found in forest and grey shrubland, usually below 900m. Adults on the wing in September–March. Larvae feed on native *Clematis* plants (vines) including *C. afoliata* and *C. marata*. **REMARKS** Abundant moth near author's home in Central Otago. Arrives every September to signify the arrival of spring. Always a welcome sight after a long, frosty winter.

GEOMETRIDAE

Large Striped Carpet ■ *Asaphodes clarata* WS 22–26mm

DESCRIPTION Endemic moth. Forewings brown with darker brown and white bands or patches spreading across wings. Hindwings orange. Variable in terms of brightness and layout of white markings on forewings. DISTRIBUTION Occurs on tussock-covered slopes. Widespread throughout the South Island and central-lower North

Island. HABITS AND HABITATS Day-flying moth that is on the wing in November–March. Also comes to light at night. Occupies open patches among tussock grassland on hills and mountains. Larvae feed on leaves of buttercups *Ranunculus* spp. Adult moths have been seen visiting flowers of the Needle-leaved Mountain Daisy *Celmisia laricifolia* and *Veronica pinguifolia*, and possibly pollinate these plants.

Aponotoreas anthracias ■ WS 22–24mm

DESCRIPTION Beautiful small, day-flying endemic moth. Characterized by stunning black and light grey zigzag patterns across forewings. This patterning enables it to be well camouflaged when resting on rocks. DISTRIBUTION Widespread in alpine and subalpine

areas of the South Island, particularly Canterbury, Otago and Southland. Only commonly seen in mountain ranges. HABITS AND HABITATS Host plants are *Dracophyllum* shrubs, which often grow in its mountainous habitat. Often seen flying by day over warmer months of year (late September–March) on South Island mountain slopes.

◼ GEOMETRIDAE ◼

Alpine Grassland Orange ◼ *Aponotoreas insignis* WS 30–34mm 🟢

DESCRIPTION Stunning endemic day-flying moth. Forewings red-brown to whitish-brown, with one white streak crossing near tip of each forewing and another diagonal streak (with thinner, fainter streak above forming teardrop shape) through centre. Lovely pattern of white streaks on wings fairly consistent between individuals, but can be more indistinct in some than others. Hindwings orange in males, without clear markings. Females much smaller and paler than males, with pale greyish forewings and light sandy-yellow hindwings. **DISTRIBUTION** Widespread, but patchy in occurrence through alpine and subalpine tussock grassland habitats across the South Island from Canterbury to Southland. Quite widespread in the mountains of Otago. **HABITS AND HABITATS** Adults primarily found on the wing in January–March. Prefers to inhabit tussockland on mountainsides of the South Island. Larvae feed on species of tussock grass, such as those in the genera *Chionochloa* and *Poa*.

Riverbed Triangle ◼ *Arctesthes catapyrrha* WS 20–24mm 🟢

DESCRIPTION Endemic, intricately patterned little day-flying moth. Adorned with curved bands across forewings in various shades of brown and light grey. Additional thinner bands in white, dark brown or black. Hindwings often orange with darker bands. **DISTRIBUTION** Widespread in the South Island, generally east of the main divide. Found in open areas from coastal to alpine locations. Prefers stony habitat such as shingle riverbeds, lake or creek edges, and stony fields. **HABITS AND HABITATS** Day flying in October–March. Larvae feed on wide range of low herbs including plantains *Plantago* spp. and the Golden Scabweed *Raoulia australis*. Adults known to pollinate the Turf Mat Daisy *R. subsericea*. **REMARKS** There are three other species in the genus with more limited distributions. For example, *A. avatar* is only known from the Denniston Plateau and nearby Mount Rochfort in the Buller District of the South Island and has a threat status of Threatened–Nationally Critical under the Aotearoa New Zealand Threat Classification System.

▪ GEOMETRIDAE ▪

Green Coprosma Carpet ▪ *Austrocidaria callichlora* WS 24–34mm (e)

DESCRIPTION Endemic. Forewings green with jagged horizontal thick or thin bands in brown, white, black and various shades of green or yellow. Hindwings pale brown or whitish. **DISTRIBUTION** Widespread throughout the North and South Islands, although appears to be more common in the latter. **HABITS AND HABITATS** Larvae feed on *Coprosma* spp. Thus, can be common in native forests and shrubland, as well as gullies with *Coprosma* on farmland, or among tussock grassland. **REMARKS** Individuals from Fiordland can sometimes be spectacularly coloured and marked, such as the one pictured here with a heavy blue-green component on forewings.

Naseby, Central Otago.

Sinbad Gully, Fiordland National Park

Austrocidaria cedrinodes ▪ WS 32–36mm (e)

DESCRIPTION Endemic. Forewings various shades of brown, often with numerous white, black and brown horizontal wavy or straight lines and bands. Can be similar to *Hydriomena hemizona*, but male *A. cedrinodes* has antennal pectinations, whereas *H. hemizona* does not. **DISTRIBUTION** Central North Island and South Island. Widespread but sparse. Also present on Rakiura/Stewart Island. **HABITS AND HABITATS** Larvae feed on *Coprosma* spp. On the wing in November–March. Attracted to light.

GEOMETRIDAE

Barred Coprosma Carpet ■ *Austrocidaria gobiata* WS 25–35mm

DESCRIPTION Endemic. Forewings various shades of light yellow-brown, often with numerous white, black and brown horizontal wavy or straight lines and bands. Hindwings similar to forewings. Diagnostic dark streak at forewing apex distinguishes it from some other Geometridae.

DISTRIBUTION Widespread throughout the North and South Islands, although appears to be rare in heavily forested regions, such as Fiordland and Kahurangi National Park. **HABITS AND HABITATS** Associated with small-leaved *Coprosma* species such as *C. propinqua*. Thus, common in shrubland, forest edges, gullies with *Coprosma* on farmland, or shrubland among tussock grassland. **REMARKS** There are two very similar species, *A. bipartita* and *A. anguligera*, both of which can be difficult to distinguish from *A. gobiata*. *A. bipartita* is fairly common in northern Aotearoa New Zealand, has a dark, tooth-like lobe on postmedian line, and is usually darker, with an apical streak only vaguely indicated. *A. anguligera* cannot be distinguished from *A. gobiata* without dissection.

Dark Coprosma Carpet ■ *Austrocidaria similata* WS 24–30mm

DESCRIPTION Endemic. Forewings green with jagged horizontal thick or thin bands in brown, white, black and various shades of green. In addition, sometimes lime-green or yellow bands. Can be distinguished from the Green Coprosma Carpet (opposite) based on two paired, large, white oval-shaped marks in centre of forewings on either side of abdomen. Hindwings pale brown or whitish. **DISTRIBUTION** Widespread throughout the North and South Islands. Also found on Auckland Islands, Campbell Island, Snares Islands, the Chatham Islands, Rakiura/Stewart Island and Codfish Island. **HABITS AND HABITATS** Larvae feed on *Coprosma* spp., thus common in forest, shrubland and other habitats containing *Coprosma*.

◾ GEOMETRIDAE ◾

Silver Fern Looper ◾ *Chalastra aristarcha* WS 35–37mm

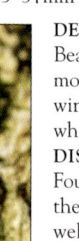

DESCRIPTION Beautiful endemic moth with distinctive wing shape and bright white markings. **DISTRIBUTION** Found throughout the North Island in well-forested habitats. **HABITS AND HABITATS** Inhabits native forest where its host plant, the Silver Fern *Cyathea dealbata*, occurs. Larvae feed on the plant during spring. Pupation occurs among leaf litter or moss on the ground. Adult moths mostly on the wing in October–April. They fly at night or can be disturbed from Silver Ferns by day. Attracted to climbing Rātā *Metrosideros perforata* when in flower. Adults come to light.

Pale Fern Looper ◾ *Chalastra pellurgata* WS 32–36mm

DESCRIPTION Endemic. Very variable in appearance. Sexes vary. Females rich golden-brown and have four clear dark lines across forewings. In male (pictured) these lines (apart from the second one) are only clear near costa and males are pale orange to dark brown, sometimes with purplish patches. **DISTRIBUTION** Throughout the country in native forest. Rare or absent in dry regions such as Central Otago and the Mackenzie Country. **HABITS AND HABITATS** Inhabits native forest. Larvae feed on various ferns, including low-growing species, tree ferns and the Silver Fern *Cyathea dealbata*. Pupates by forming thin cocoon on the soil. Adults on the wing throughout the year. By day adult moths can be seen resting on dead fern fronds. They become active from dusk and are attracted to light.

GEOMETRIDAE

Australian Pug Moth ■ *'Chloroclystis' filata* WS 20–25mm

DESCRIPTION Probably self-introduced from Australia in about 1960. One of the most commonly encountered moths in Aotearoa New Zealand, and can be distinguished from other pug moths by dark band across hindwings, visible when the moth is at rest. In some individuals there are large white patches on forewings. **DISTRIBUTION** Common and widespread throughout the North and South Islands, Rakiura/Stewart Island and Chatham Islands. **HABITS AND HABITATS** Larvae feed on flowers of a variety of flowering plants. In New Zealand they include *Senecio* spp. and Gorse *Ulex europaeus*. On the wing throughout the year and commonly attracted to light.

Kawakawa Looper ■ *Cleora scriptaria* WS 30–55mm

DESCRIPTION Endemic and well known for considerable variation in adult size, colour and wing patterns. Both male and female have spot in middle of forewing that can be black, white or cream. Scalloping on termen of both forewings and hindwings can help with recognizing species. **DISTRIBUTION** Reasonably common throughout the country, including on Rakiura/Stewart Island. **HABITS AND HABITATS** Usually found in forested areas near or on Kawakawa *Piper excelsum* and other host plants. Larvae make distinctive holes as they feed at night. They can also be found on the Wineberry *Aristotelia serrata*, Horopito *Pseudowintera* sp., Ramarama *Lophomyrtus* sp., Akeake *Dodonaea viscosa* and introduced Feijoas *Feijoa sellowiana*. Adults hide on tree trunks or among leaf litter by day. They fly at night and are attracted to light.

▪ GEOMETRIDAE ▪

Red-spotted Delicate ▪ *Epicyme rubropunctaria* WS 25mm

DESCRIPTION The only member of the genus *Epicyme*. Yellow-white, orange-white or pinkish-white moth, named for distinctive red spots or patches, which it sometimes has on forewings. Also adorned with black spots and crenulate bands in dark grey or black across wings. Row of well-spaced black spots along termen of both forewings' hindwings. **DISTRIBUTION** Widespread throughout the North and South Island, Rakiura/Stewart Island and the Chatham Islands. Also found in eastern, southeastern and southwestern Australia, including Tasmania. **HABITS AND HABITATS** Larvae have been recorded feeding on plants in the genera *Haloragis*, *Gaultheria* (wintergreens) and *Geranium*. On the wing year round, but less commonly seen in winter, and most commonly seen in mid-spring.

Declana niveata ▪ WS 30–37mm e

DESCRIPTION Endemic. Whitish, cream or light grey in colour. Similar in appearance to some examples of the very variable Forest Semilooper (opposite). However, unlike that species, this one has pale, unmarked hindwings and no antennal pectinations on male, and antemedian line runs obliquely to near costa, then fades, whereas in the Forest Semilooper it either fades well before the costa or there is a second part from the costa that meets it in a 'V' shape. **DISTRIBUTION** Widespread throughout most of Aotearoa New Zealand but infrequently encountered. **HABITS AND HABITATS** Larvae feed on *Hoheria* spp. Adults primarily on the wing in warmer months of the year (October–March), and attracted to light.

GEOMETRIDAE

Forest Semilooper ■ *Declana floccosa* WS 27–35mm

DESCRIPTION Endemic moth, very beautiful and highly variable in colour and pattern. See *D. niveata* (opposite) for how to distinguish it from that species. **DISTRIBUTION** Widespread and common throughout the North and South Islands. Also found on the Chatham Islands and Rakiura/Stewart Island. **HABITS AND HABITATS** Larvae polyphagous and feed on a wide range of native and exotic broadleaved and coniferous shrubs and trees. Native host plants include Pōhuehue *Muehlenbeckia australis*, Kauri *Agathis australis*, korokio *Corokia* spp., tutu *Coriaria* spp., colicwood *Myrsine* spp., beech *Fuscospora* spp., *Dacrydium* spp. and Putaputaweta *Carpodetus serratus*. Exotic hosts include pine trees (*Pinus radiata* and other *Pinus* spp.), Douglas Fir *Pseudotsuga menziesii*, larches *Larix* spp. and *Eucalyptus* spp. Due to its wide range of host plants and ability to utilize exotic vegetation, such as pines, this species is highly adaptable to modified environments and lives in a wide range of habitats. One of the most commonly seen moths in Aotearoa New Zealand and readily comes to light.

GEOMETRIDAE

Lawyer Pug ■ *Elvia glaucata* WS 24–28mm

DESCRIPTION One of Aotearoa New Zealand's most beautiful moths. The alien-like *E. glaucata* occurs in a stunning array of colour combinations, often mixing shades of green, grey, white, blue, light yellow, and sometimes orange or purple. Strong fold in forewing visible when moth is at rest. **DISTRIBUTION** Widespread and common throughout the North and South Islands. Also found on the Chatham Islands and Rakiura/Stewart Island. **HABITS AND HABITATS** Green or pinkish larvae feed on the Bush Lawyer *Rubus cissoides*. Readily comes to light in forests or shrubland. On the wing year round, but less commonly seen in winter.

◾ GEOMETRIDAE ◾

Lacebark Looper ◾ *Epiphryne undosata* WS 23–27mm (e)

DESCRIPTION Endemic. Adult moths can be very variable in the intensity of their colours and markings. Some individuals very plain yellowish-white with only faint markings, while others bright yellow with strong black, dark-grey or purplish-red bands. **DISTRIBUTION** Widespread on the North and South Islands. **HABITS AND HABITATS** Inhabits native forest. Larvae feed on native species of Malvaceae including *Hoheria* spp. and probably *Plagianthus regius*. They pupate among dead leaves in a silk cocoon.

Adults feed from flowers and may assist with pollination of *Dracophyllum acerosum* and *Veronica salicifolia*. They have been seen on the wing year round, but are most commonly sighted in November–February. Lacebark Loopers can be found in great numbers around forest margins with a lot of lacebark trees *Hoheria* spp. Adults nocturnal and attracted to light.

Cabbage Tree Moth ◾ *Epiphryne verriculata* WS 40mm (e)

DESCRIPTION Distinctive endemic moth with numerous parallel horizontal light brown lines across both forewings and hindwings. **DISTRIBUTION** Widespread on the North and South Islands, and Rakiura/Stewart Island. In the South Island, most common in coastal areas and rare in inland regions like Central Otago. **HABITS AND HABITATS** Larvae feed on the cabbage tree *Cordyline australis* and all other native species of *Cordyline*. Wings patterned to camouflage adult moth against dead cabbage-tree leaves, where it often rests. Favoured habitats are wetlands and native forest. Also found in urban areas, as cabbage trees are often grown in town gardens and public spaces where people live. Adults seen from spring to late summer (October–May) and come to light.

GEOMETRIDAE

Brown Evening Moth & Lesser Brown Evening Moth
■ *Gellonia dejectaria* & *G. pannularia* WS 40–50mm 🟢

DESCRIPTION *Gellonia* is an endemic genus of large, docile moths. The Brown Evening Moth is larger on average but the two species are very similar in appearance; only the Lesser Brown Evening Moth has thin black line on underside of both forewings next to thicker brown band. **DISTRIBUTION** Both species widespread throughout the North, South and Rakiura/Stewart Islands. **HABITS AND HABITATS** Occur in native forests and

Brown Evening Moth Gellonia dejectaria

shrubland. On the wing throughout the year, but most common in summer. Both feed on a range of trees, shrubs and climbers. For one or both species this includes: tutu *Coriaria* spp., māhoe *Melicytus ramiflorus*, Houpara *Pseudopanax lessonii*, Supplejack *Ripogonum scandens* and Bush Lawyer *Rubus cissoides*. They rest camouflaged on tree trunks by day and fly at night. Both species attracted to light.

Left and right: Lesser Brown Evening Moth Gellonia pannularia

Dark-banded Carpet Moth ■ *'Hydriomena' deltoidata* WS 32–36mm 🟢

DESCRIPTION Endemic moth with a beautiful mixture of transverse crenulate bands and circular shapes in white, various browns and greys. Light orange hindwings. Black, oval-shaped discal spots on forewings. Variable in forewing pattern; male forewing narrower than that of female. **DISTRIBUTION** Widespread and locally common throughout the North, South, Chatham and Rakiura/Stewart Islands. Rare in Northland and Auckland. **HABITS AND HABITATS** Larvae feed on plantains and probably other herbaceous plants. Adults visit and may pollinate *Dracophyllum acerosum* and *Leptospermum* spp. On the wing in November–March and attracted to light. **REMARKS** Does not belong in the Hydriomena or even in the tribe Hydriomenini; it is in the Xanthorhoini and the genus placement needs revising.

◾ GEOMETRIDAE ◾

New Zealand Looper ◾ *Epyaxa rosearia* & other *Epyaxa* spp.
WS 22–26mm (e)

DESCRIPTION There are nine species in the genus *Epyaxa*. Three are endemic to Aotearoa New Zealand and the other six occur in Australia. The most common and widespread species in New Zealand is the New Zealand Looper. In this species, adults are varied in appearance, being pinkish, grey, brownish or olive-green. Some New Zealand Loopers are very similar to lightly marked examples of the Barred Pink Carpet (p. 75), but the latter usually has a rather distinct, pale-scaled vein running longitudinally through middle of forewing that forks in centre of wing, looking like an elongated 'F'.

DISTRIBUTION Widespread in the North and South Islands, Chatham Islands and Rakiura/Stewart Island. *E. lucidata* widespread in the North Island and from Canterbury northwards in the South Island. *E. venipunctata* widespread but uncommonly encountered.

HABITS AND HABITATS Larvae of the New Zealand Looper feed on a variety of herbaceous plants, including Watercress *Nasturtium officinale*, *Trifolium* spp., *Plantago* spp., *Tropaeolum* spp. and *Gunnera* spp. *Epyaxa venipunctata* has been associated with Kokihi *Tetragonia implexicoma*. All three of these species are on the wing year round and come to light.

Some variation within the New Zealand Looper Epyaxa rosearia

Epyaxa lucidata

Epyaxa venipunctata

GEOMETRIDAE

Helastia spp. ■ WS 18–30mm

DESCRIPTION *Helastia* species are silvery-white looper moths that occur in an array of subtly beautiful colours and patterns. There are 18 described species in this endemic genus. Some, such as *H. cinerearia*, are common and widespread, whereas others, like *H. expolita*, are quite rare. Six species are profiled below. **DISTRIBUTION** *H. cinerearia* found in the North, South, Chatham and Rakiura/Stewart Islands; *H. corcularia* across the South, Chatham and Rakiura/Stewart Islands; *H. alba* in the South Island and central-lower North Island; *H. christinae* in coastal Otago, Central Otago, Mackenzie Country and Otago Lakes areas; *H. cryptica* from mid-Canterbury to Southland; *H. expolita* in Buller, Marlborough and Canterbury regions. **HABITS AND HABITATS** *H. cinerearia* can be found in urban gardens, forest and subalpine habitats. Adult moths feed from and probably pollinate *Veronica salicifolia*, *Hoheria lyallii* and *Leptospermum* spp. *H. corcularia* inhabits a wide variety of habitats. *H. alba* inhabits beech forests, podocarp forests and shrubland. *H. christinae* occurs in subalpine to lowland areas, frequenting riverbanks and streamsides. *H. expolita* favours tussock grassland in alpine and subalpine zones. Life history for all species poorly known, but most seem to be associated with mosses and/or herbs, possibly also lichens. All are nocturnal and come to light.

Helastia cinerearia

Helastia corcularia

Helastia alba

Helastia cryptica

Helastia expolita

Helastia christinae

GEOMETRIDAE

Small Hooked-tip Looper ▪ *Homodotis megaspilata*, & *H. falcata* WS 22–26mm e

Small Hooked-tip Looper Homodotis megaspilata

DESCRIPTION The genus *Homodotis* contains three endemic species. The two most common are profiled: the Small Hooked-tip Looper *H. megaspilata*, and *H. falcata*. The Small Hooked-tip Looper is very variable in appearance but can be distinguished from similar species, as all variations have forewings with blunt, hook-shaped tips. Small Hooked-tip Loopers might possibly be confused with *H. falcata*, as their range overlaps from Dunedin south. However, *H. falcata* is a larger moth but with less strongly hooked forewings. **DISTRIBUTION** The Small Hooked-tip Looper is widespread country wide; *H. falcata* is only found in the southern South Island (Otago and Southland). Both occur on Rakiura/Stewart Island. **HABITS AND HABITATS** The Small Hooked-tip Looper is found in native forest, shrubland, coastal areas and domestic gardens. Larvae of both species feed on dead leaves of native plants; those of Hangehange *Geniostoma ligustrifolium* have been confirmed as hosting the Small Hooked-tip Looper. Adult moths visit flowers of *Melicytus ramiflorus*, *Olearia virgata* and *Leptospermum* spp. Both species nocturnal and attracted to light.

Homodotis falcata

GEOMETRIDAE

'Hydriomena' rixata ■ WS 26–28mm

DESCRIPTION
Gorgeous endemic moth with brilliant green, white, grey and blue lines and bands across forewings. There are also quite dull brownish forms with little green in the southern South Island, illustrated in the photograph on the left. Unlikely to be confused with any other species.
DISTRIBUTION
Widespread throughout the North and South Islands. **HABITS AND HABITATS** Found in native forest edges and clearings, shrubland and open areas. Larvae feed on plantains and willowherbs *Epilobium* spp. Adults known to visit and may pollinate *Dracophyllum acerosum*. *Hydriomena rixata* on the wing in September–March and attracted to light.

GEOMETRIDAE

Ipana griseata ◼ 36–40mm e

DESCRIPTION Attractive dark grey endemic moth. Adult distinguished from other related *Ipana* and *Declana* species with pectinate antennae by its 'fluffier', more finely scaled thorax and pale forewing band, which runs across *before* middle of wing. Adults larger and darker grey than grey forms of the Forest Semilooper (p. 51) and *Declana nigrosparsa* (p. 158). **DISTRIBUTION** Widespread throughout the South Island and Rakiura/Stewart Island. Also found in the North Island but very patchy in occurrence. **HABITS AND HABITATS** Larvae feed on foliage and shoots of mistletoes, that is Green Mistletoe *Ileostylus micranthus*, Scarlet Mistletoe *Peraxilla colensoi*, Red Mistletoe *P. tetrapetala* and White Mistletoe *Tupeia antarctica*. On the wing throughout the year and attracted to light. **REMARKS** *I. griseata* classified as At Risk, predicted decline 10–70 per cent (Hoare et al. 2017). This reflects the observed reduction (extinction, in some places) of its mistletoe hosts, particularly in the North Island, and this is usually attributed to possum browsing.

North Island Zebra Moth ◼ *Ipana atronivea* WS 36–54mm

DESCRIPTION Exquisite moth endemic to the North Island. It is large with a distinctive contrasting pattern of white and black. Unlikely to be confused with any other moth in Aotearoa New Zealand. The most similar species is the South Island Zebra Moth (p. 60), but the two species occur on separate islands. Differences in pattern between them are consistent. **DISTRIBUTION** Widespread throughout the North Island. **HABITS AND HABITATS** Occurs in native forests and shrubland from the coast to 1,200–1,300m. Larva resembles a twig of its host plant, the Five Finger *Pseudopanax arboreus*, when covered in lichens or a fungus (*Septobasidium*) that hosts scale insects. Other host plants include the Mountain Five Finger *P. colensoi*, Lancewood *P. crassifolius*, *Raukaua edgerleyi* and *R. simplex*. Adults on the wing in September–April and attracted to light. **REMARKS** Until 2023, known as *Declana atronivea*, but recent taxonomic work has identified that it (and several other related species) should be placed in a separate genus, *Ipana*.

■ GEOMETRIDAE ■

South Island Zebra Moth ■ *Ipana egregia* WS 40–54mm

DESCRIPTION Utterly gorgeous moth endemic to southern Aotearoa New Zealand. Large, with black and white head and thorax, and bold and sharply defined forewing pattern. **DISTRIBUTION** Widespread throughout forested parts of the South Island and Rakiura/Stewart Island. Appears to be more common in high-rainfall areas and absent from dry regions such as Central Otago. **HABITS AND HABITATS** Found in native forests and shrubland from sea level to low-alpine zone. Larva looks like a twig of its host plant, the Five Finger *Pseudopanax arboreus*, when covered in lichens or scale insects. Other host plants include the Mountain Five Finger *P. colensoi* and *Raukaua simplex*. Adults on the wing throughout most of the year, except mid-winter, and attracted to light. **REMARKS** The two sister genera *Ipana* (10 described species) and *Declana* (five species) are together informally referred to as the 'declanoids'. The North Island Zebra Moth *Ipana atronivea* (p. 59) and *I. egregia* are the only two declanoids to have diverged and formed a sister-species pair between the North and South/Stewart Islands.

Ipana junctilinea ■ WS 34–43mm

DESCRIPTION Endemic moth with a prominent dark crest (height 3mm) on thorax. Antemedian and postmedian lines are usually doubled and may be joined to form an 'X' across the wing. Populations in inland grey shrubland of the South Island (for instance in the Mackenzie Country) tend to be predominantly grey not brown. **DISTRIBUTION** Widespread across the North and South Islands, and Rakiura/Stewart Island. **HABITS AND HABITATS** Occupies a range of habitats, including forests and particularly shrubland. Larvae highly polyphagous and feed on plants in many different families, including a wide range of endemic shrubs and trees; also on *Muehlenbeckia* vines and introduced conifers. Mainly on the wing in October–March, but can be encountered in any month. Comes to light. Prominent 3mm-high crest on thorax resembles stump of a broken off twig and adds to the disruptive appearance of the adult moth as it clings to a shrub with its forewings firmly wrapped around a twig.

▪ GEOMETRIDAE ▪

Spotted Mānuka Looper ▪ *Ipana leptomera* WS 36–40mm 🟢

DESCRIPTION Endemic moth. Males fawn and females grey. Forewing costa has a row of herringbone-shaped strigulae (marks that look like this on the costa: >>>>>). This distinguishes the Spotted Mānuka Looper from other *Ipana* and *Declana* species.
DISTRIBUTION Widespread across the North, South and Rakiura/Stewart Islands.
HABITS AND HABITATS Range of habitats including forests and shrubland. Larvae highly polyphagous, feeding on plants in many families, including a wide range of endemic shrubs and trees, such as conifers, mānuka *Leptospermum* spp. and the mistletoe *Ileostylus micranthus*. Mainly on the wing in September–April but may be encountered in any month. Comes to light.

Zigzag Fern Looper ▪ *Ischalis fortinata* WS 34–40mm 🟢

DESCRIPTION Endemic, distinctive pinkish-orange looper moth with prominent brown or black zigzag lines across centre of forewings. **DISTRIBUTION** Widespread on the South Island and Rakiura/Stewart Island. Also occurs throughout much of the North Island but less common and sparsely distributed. **HABITS AND HABITATS** Found in native forests and shrubland. Larval hosts are the shield ferns *Polystichum vestitum* and *P. richardii*. Larvae feed all year round. Eggs laid on leaves of host species. Adult moths on the wing primarily in October–March.

◾ GEOMETRIDAE ◾

Striped Fern Looper ◾ *Ischalis gallaria* WS 34–36mm (e)

DESCRIPTION Master of camouflage, mimicking a dead leaf on the forest floor. Endemic. Somewhat similar to the Zigzag Fern Looper (p. 61) but much more variable in colour and markings. Also, postmedian line across forewings is straight in this species, rather than zigzag shaped, and male hindwing is not scalloped. **DISTRIBUTION** Widespread on the North, South and Rakiura/Stewart Islands. Rare in inland, central South Island. **HABITS AND HABITATS** Larval hosts are thought to be ferns, including *Blechnum* spp., *Microsorum* spp. and the Gully Fern *Pneumatopteris pennigera*. Shape and colouring of wings look like a fallen/dead leaf on the forest floor. Adult moths on the wing primarily in October–April.

Oblique-waved Fern Looper ◾ *Ischalis variabilis* WS 36–42mm

DESCRIPTION Beautiful endemic forest-dwelling species. Both sexes have four discal spots. Somewhat similar to the Zigzag Fern Looper (p. 61), but a light yellow-white colour v orange, and often covered in black specks and spots. Stripes across forewings also less sharply angled than in the Zigzag Fern Looper. **DISTRIBUTION** Widespread in the North Island. Also found on western side of the South Island. **HABITS AND HABITATS** Larvae have been recorded feeding on tree ferns, including *Cyathea smithii*, *C. dealbata* and *Dicksonia squarrosa*. Adults have been recorded year round, but main flight period is September–April.

◾ GEOMETRIDAE ◾

Notoreas paradelpha ◾ WS 21–23mm

DESCRIPTION Stunning endemic day-flying moth of alpine habitats. **DISTRIBUTION** Widespread in alpine areas of the South Island above the tree line from Nelson region to Southland. **HABITS AND HABITATS** Lives in alpine habitats and has been found among tussock grassland and alpine herb fields. Host plants for larvae are endemic species in the genera *Kelleria* and *Pimelea*, including *Pimelea oreophila*. Main flight period is November–January. **REMARKS** There are more than 20 day-flying, brightly coloured endemic moth species in the genus *Notoreas*. There are alpine and coastal species, some of which are undescribed and facing threats from habitat loss.

Flax Window-maker ◾ *Orthoclydon praefectata* WS 38–45mm

DESCRIPTION Glossy-white endemic moth. Larvae leave distinctive oval or rectangle-shaped windows in flax leaves when feeding (see below right). Male moth pale brown; female glossy-white. Small individuals could be confused with the Native Cranberry Moth (p. 64), but note difference in shape of hindwings at termen. **DISTRIBUTION** Widespread on the North, South and Rakiura/Stewart Islands. **HABITS AND HABITATS** Occupies wetlands, shrubland and forest edges. Food plant is the New Zealand native flax *Phormium tenax*. On the wing throughout the year. Attracted to light. **REMARKS** The endemic genus *Orthoclydon* also contains the Snowberry Yellow (p. 162) and the very rare *O. pseudostinaria*, which is classified as Critically Endangered by the Department of Conservation.

GEOMETRIDAE

Native Cranberry Moth ■ *Poecilasthena pulchraria* WS 20–25mm

DESCRIPTION Delicate whitish moth with subtle grey, red, green and blue tones. Native to both Aotearoa New Zealand and Australia. Can be recognized by numerous white-grey, white-green or white-blue, wavy lines across both forewings and hindwings. Angled hindwing also distinctive and helps distinguish it from the Kānuka Looper (p. 68). Sometimes a reddish or brown edge along costa of forewings. **DISTRIBUTION** Widespread in Australia and New Zealand. **HABITS AND HABITATS** Found around native forest and shrubland. Adults mainly recorded in October–May. Flies at night and comes to light. Green larvae feed on flowers and leaves of the Minigimingi *Leucopogon fasciculatus* or mānuka *Leptospermum* spp.

Orange Underwing ■ *Paranotoreas brephosata* WS 24–28mm

DESCRIPTION Endemic day-flying moth known for its striking orange colouration. Forewings dark grey. Hindwings bright orange, with 2–4 distinct wavy black transverse lines that are generally rather narrow. Termen broadly bordered with black. **DISTRIBUTION** Widespread on the South Island. Also present in the central and lower North Island. **HABITS AND HABITATS** Day flying in dry-stony habitats, along creeks, wetlands, by lakes and among tussock grassland. Larvae feed on willowherbs *Epilobium* spp. On the wing in November–April.

GEOMETRIDAE

Creekbed Orange Underwing ■ *Paranotoreas zopyra* WS 24–28mm e

DESCRIPTION Endemic day-flying moth known for its striking bright orange hindwings. Forewings dark grey or blueish, sometimes rather plain, or charmingly marked with orange bands or bright blues. Hindwings bright orange, with 2–4 distinct wavy black, transverse lines that are generally rather narrow. Termen of hindwings either has no black or only a thin black border, whereas the similar Orange Underwing (opposite) has a thicker black band. **DISTRIBUTION** Widespread in the South Island. Mostly found in mountains in subalpine and alpine zones. Also occurs in the central and lower North Island. **HABITS AND HABITATS** Day flying in dry-stony habitats, along creeks, in wetlands, by lakes and among tussock grassland. Larvae feed on willowherbs *Epilobium* spp. Common in montane to high-alpine settings but rare at sea level. **REMARKS** There are five described species of day-flying moth in the *Paranotoreas* genus, as well as an undescribed species that was discovered in a herb field on the summit of Mount Herbert on Banks Peninsula by Brian Patrick in 2005.

GEOMETRIDAE

Pug moths ■ *Pasiphila* spp. WS 15–30mm

DESCRIPTION *Pasiphila* is a genus of more than 45 pug moth species found nearly worldwide. Aotearoa New Zealand is a global hotspot for it, with more than 30 endemic species, including some undescribed ones. Pug moths can be very difficult to identify to species level without expert assistance, but some species are highly distinctive, such as the stunning Mottled Forest Pug *P. lichenodes*. The male Pome Looper *P. testulata* has a distinctive raised distortion of front edge of forewing, which is a pocket holding scent-scales. **DISTRIBUTION** *Pasiphila* moths occur throughout New Zealand, including the North, South and Rakiura/Stewart Islands. One species, *P. nebulosa*, on the Auckland Islands. More than 30 species on the South Island, while the North Island holds at least 20. Some species widespread and reasonably common throughout the country, such as *P. inductata*, the Emerald Pug Moth *P. muscosata*, and *P. lunata*, *P. bilineolata*, *P. plinthina* and *P. sandycias*. The Pome Looper is widespread and also found in Australia, and on Norfolk Island and the Chatham and Kermadec Islands. Some species, such as *P. aristias*, *P. cotinaea*, *P. heighwayi*, *P. furva* and *P. charybdis* rarely encountered, with few known sites. **HABITS AND HABITATS** Nocturnal moths that come to light. Found in a wide range of habitats, particularly native forests, shrubland and grassland. Larval host plants varied, including many native trees, shrubs and vines. For example, the Emerald Pug Moth feeds on *Muehlenbeckia* spp.; *P. lunata*, *P. rubella*, *P. malachita*, *P. charybdis*, *P. heighwayi*, and *P. bilineolata* on hebes *Veronica* spp.; *P. melochlora* on *Carmichaelia* spp.; *P. magnimaculata* on *Gaultheria crassa*; *P. humilis* on *Dracophyllum* spp. and *Epacris* spp.; *P. urticae* on the Tree Nettle *Urtica ferox*; *P. nereis* on *Celmisia* spp. and *Craspedia* spp., and *P. cotinaea* on *Olearia* spp. Adult *Pasiphila* moths probably pollinate a wide range of native plants, including those in the genera *Leptospermum*, *Olearia*, *Dracophyllum* and *Veronica*.

Pasiphila inductata

Pome Looper Pasiphila testulata

Pasiphila rubella

Pasiphila sphragitis

GEOMETRIDAE

Pasiphila halianthes

Pasiphila fumipalpata

Pasiphila nereis

Mountain Pug Pasiphila magnimaculata

Pasiphila lunata

Emerald Pug Moth Pasiphila muscosata

Green Broom Pug Pasiphila melochlora

Pasiphila punicea

Pasiphila bilineolata

Pasiphila sandycias

Pasiphila dryas

Mottled Forest Pug Pasiphila lichenodes

GEOMETRIDAE

Apple Looper ■ *Phrissogonus laticostata* WS 15mm

DESCRIPTION Self-introduced resident (native to Australia). Mottled grey with brown bands across forewings. Moths roughly triangular when at rest. Male moths have tufts on leading edge of forewing. **DISTRIBUTION** Australia, New Caledonia and Aotearoa New Zealand. In the latter widespread in the North Island, but in the

South Island only common from Canterbury northwards. Also recorded on the Chatham Islands. **HABITS AND HABITATS** Can damage fruits in apple orchards and vineyards. Larvae are polyphagous and predominantly feed on flowers. Attracted to light. Adults found mainly in January–April.

Kānuka Looper ■ *Poecilasthena schistaria* WS 20–25mm

DESCRIPTION Attractive endemic moth common in many shrublands. Forewings pale white-brown to dull purplish-brown. Double dashes along termen distinguish it from the Red-spotted Delicate (p. 50), which has well-spaced dots. **DISTRIBUTION** All around the North, South and Rakiura/Stewart Islands. **HABITS AND HABITATS** Occurs around native forest and shrubland, and also found in towns, cities and farmland. Adults recorded throughout the year, but most commonly in September–April. Flies by night and comes to light. Green larvae feed on mānuka *Leptospermum* spp. and kānuka *Kunzea* spp. Pupation occurs in a slight cocoon.

■ GEOMETRIDAE ■

Pseudocoremia fenerata ■ WS 27–34mm 🟢

DESCRIPTION Endemic whitish-grey looper moth. Most similar to *P. rudisata* (p. 71) and *P. indistincta* (below) but large hindwings peeking out from underneath forewings when the moth is at rest are characteristic of male *P. fenerata*. *P. indistincta* typically darker, being more blackish-grey or sometimes greenish-grey. **DISTRIBUTION** Throughout the North, South and Rakiura/Stewart Islands. **HABITS AND HABITATS** Occurs in and around native forest and shrubland. May also be found in town gardens and around pine plantations. Adults on the wing throughout the year, but more common in warmer months. *P. fenerata* flies at night and comes to light. Larvae utilize a range of conifers, including Kauri *Agathis australis*, Bog Pine *Dacrydium bidwillii*, Pink Pine *D. biforme*, Rimu *D. cupressinum*, Mountain Toatoa *Phyllocladus alpinus*, Tānekaha *P. trichomanoides*, and *Podocarpus totara*. Exotic hosts are cedars *Chamaecyparis lawsoniana* and *Cryptomeria japonica*, larches *Larix decidua* and *L. kempferi* and pines *Pinus* spp. **REMARKS** *Pseudocoremia* is an endemic genus of looper moths with about 30 species.

Pseudocoremia indistincta ■ WS 30–34mm 🟢

DESCRIPTION Subtly beautiful endemic looper moth. Forewings grey-brown with dark and white bands, patches and transverse lines. Most similar to *P. rudisata* (p. 71) and *P. fenerata* (above), but darker in overall colour and hindwing is yellowish or pale orange, v pale or less coloured.
DISTRIBUTION Widespread in the North and South Islands.
HABITS AND HABITATS Found around native forest and shrubland. May also occur around towns and cities. Adults on the wing mostly in November–April. Flies at night and comes to light. Larvae feed on endemic pōhuehue *Muehlenbeckia* spp. **REMARKS** Males of all *Pseudocoremia* species have strongly feathery antennae (or large pectinations).

GEOMETRIDAE

Forest Looper ■ *Pseudocoremia leucelaea* WS 30–34mm

DESCRIPTION Endemic moth that occurs in an array of colour forms. Forewings have bright white or yellowish-white thick band with a few gaps in it across forewings and a thin white wavy line along dorsum. Most individuals predominantly brown or yellow-brown, with patches and bands in white, but there is also a white, grey and black form. The latter can sometimes be confused with the Celery Pine Looper *P. monacha* but this has a more even covering of white and black patches, and white band across forewings is thinner.
DISTRIBUTION Widespread in the North and South Islands. **HABITS AND HABITATS** Found around shrubland and native forest. May also occur around towns and cities. Adults on the wing year round but most common in November–May. Flies at night and comes to light. Larvae feed on various conifers, both native and introduced. Native conifers include Kauri *Agathis australis*, Mountain Toatoa *Phyllocladus alpinus*, Tānekaha *P. trichomanoides* and *Podocarpus totara*.

Male

Female

Pseudocoremia lupinata ■ WS 32–37mm

DESCRIPTION Subtly beautiful endemic moth. Forewings pale dull pinkish-brown. Hindwings pale ochreous. **DISTRIBUTION** Widespread in the North and South Islands, but infrequent and rare in the northern North Island. **HABITS AND HABITATS** Favoured habitat is kānuka *Kunzea* spp. shrubland, as this is the larval host. Both the larvae and adults are nocturnal. Adult moths commonly on the wing in December–April, and attracted to light.

GEOMETRIDAE

Brown Forest Flash ■ *Pseudocoremia productata* WS 30–35mm

DESCRIPTION Magnificent and variable endemic moth. Forewings a mixture of browns and yellows with white transverse lines. There are lighter and darker forms (as shown) and there is considerable variation between individuals. **DISTRIBUTION** Widespread in the North, South and Rakiura/Stewart Islands. **HABITS AND HABITATS** Occurs in shrubland and forest. Larvae feed on foliage of shrubs and trees including Miro *Prumnopitys ferruginea*, Southern Rātā *Metrosideros umbellata* and tutu *Coriaria* spp. Both larvae and adults are nocturnal. Adult moths on the wing throughout the year and attracted to light.

Pseudocoremia rudisata ■ WS 34–38mm

DESCRIPTION Endemic moth with two similar subspecies: *P. r. ampla* and *P. r. rudisata*. The latter has a greater covering of darker blotches. **DISTRIBUTION** Widespread. *P. r. rudisata* predominantly found in the North Island, and *P. r. ampla* in the South Island and Rakiura/Stewart Island. **HABITS AND HABITATS** Occurs in shrubland and forest. Larvae feed on small-leaved tree daisies such as *Olearia bullata*, *O. fimbriata*, *O. fragrantissima* and *O. odorata*, as well as Rangiora *Brachyglottis repanda*. Both larvae and adults are nocturnal. Adult moths on the wing throughout the year, most commonly in spring and early summer. Attracted to light.

Left and above: Pseudocoremia rudisata ampla (*adult moth and looping caterpillar*) *Right:* Pseudocoremia rudisata rudisata

◼ GEOMETRIDAE ◼

Common Forest Looper ◼ *Pseudocoremia suavis* WS 26–34mm

DESCRIPTION Endemic moth that is highly variable in appearance. Usually, forewings grey or pale white-brown, speckled and mottled with black or dark brown, and hindwings uniform pale yellow. Females generally more contrastingly marked than males and do not have pectinate antennae. **DISTRIBUTION** Widespread and common in the North, South and Rakiura/Stewart Islands. One of the most commonly encountered endemic moths in Aotearoa New Zealand. **HABITS AND HABITATS** Occurs in a wide range of habitats, including shrubland, forests, town and city gardens, exotic plantations and rural areas. Larvae feed on a wide range of plants, including southern beech *Fuscospora* spp., podocarps and kānuka *Kunzea* spp. Adults on the European Gorse *Ulex europaeus* and a variety of exotic trees, including pines *Pinus* spp. Both larvae and adults are nocturnal. Adult moths feed on nectar of flowers and on honeydew excreted by scale insects. They may be disturbed from the litter or from resting places on trees by day. Adult moths on the wing throughout the year and attracted to light.

▪ GEOMETRIDAE ▪

Hook-tip Fern Looper ▪ *Sarisa muriferata* WS 32–36mm

DESCRIPTION *Sarisa* is an endemic monotypic moth genus with one species, the Hook-tip Fern Looper. Well known for its unusual shape and distinctive hooked wing-tips. Often a distinctive horizontal line through centre of both forewings and hindwings. **DISTRIBUTION** Widespread in the North and South Islands, and recorded from Stewart Island, Big South Cape Island, the Chatham Islands and the Auckland Islands. **HABITS AND HABITATS** Adults found in native forest. Larvae night feed on ferns, including the Hound's Tongue Fern *Microsorum pustulatum*, Leatherleaf Fern *Pyrrosia eleagnifolia* and Whekī-Ponga *Dicksonia fibrosa*. Adult moths on the wing throughout the year. Attracted to light.

Common Fern Looper & Huarau Looper ▪
Sestra flexata & *S. humeraria* WS 30–34mm

DESCRIPTION Endemic moths that can be ash-grey, fawn, orange or purplish in colour. Hindwings yellowish, without markings. Three evenly spaced short, dark markings along costa of forewings; middle one may extend as horizontal stripe across centre of moth. Species distinguishable based on wing shape along termen (base), which is more strongly curved in the Common Fern Looper. **DISTRIBUTION** Both species found throughout Aotearoa New Zealand, including the North, South, Rakiura/Stewart and Chatham Islands. **HABITS AND HABITATS** Larvae of the Common Fern Looper feed on ferns such as the Sweet Fern *Pteris macilenta*, Bracken *Pteruidium esculentum* and Mātā *Histiopteris incisa*. Adults recorded visiting flowers of mānuka *Leptospermum* spp. Larvae of the Huarau Looper also feed on ferns, probably including the Lace Fern *Paesia scaberula* and Thousand-leaved Fern *Hypolepis millefolium*. Adult moths of both species nocturnal and attracted to light. The Common Fern Looper is mostly on the wing in August–April (peak in January). The Huarau Looper is a spring moth, mostly on the wing in July–December.

Sestra humeraria

Sestra flexata

▪ GEOMETRIDAE ▪

Plantain Moth ▪ *Scopula rubraria* WS 20mm

DESCRIPTION Sun-loving, day-flying moth. Self-introduced resident, native to both Aotearoa New Zealand and Australia. A brownish moth, sometimes with an attractive pinkish hue on hindwings. Intensity of black specking varies among individuals. **DISTRIBUTION** Found throughout New Zealand as far south as Otago and in coastal regions of southeastern Australia and throughout Tasmania. **HABITS AND HABITATS** Can be abundant in open areas, pastures, grassland, weedy areas, roadsides and gardens. Larvae feed on the Narrow-leaved Plantain *Plantago lanceolata*, and have also been raised on the Sickle Alfalfa *Medicago sativa* and on the endemic plantain *Plantago spathulata*. Adult moths on the wing in October–April but most abundant in late summer. Flies actively in sunny weather and may also come to light at night. **REMARKS** *Scopula* is a huge genus of moths found worldwide, with more than 700 species; however, *S. rubraria* is the only member of the genus that resides in New Zealand.

Kāmahi Green Spindle ▪ *Tatosoma tipulata* WS 34–36mm

DESCRIPTION Stunning olive-green moth. Forewings long, narrow and slightly rounded at tips, with several undulating, irregular brown or blackish lines. Hindwings lightly marked or not marked and much paler. Like all *Tatosoma* species (spindle moths), males have distinctive thin, elongated abdomen that stretches well beyond base of wings. In females abdomen ends approximately at termen of forewings. **DISTRIBUTION** Throughout Aotearoa New Zealand in mature native forests, including on Rakiura/Stewart Island. **HABITS AND HABITATS** Inhabits mature native forest. Larvae feed on foliage of a wide range of native trees, including the Kamahi *Pterophylla racemosa*, Mountain Beech *Fuscospora cliffortioides* and Tōtara *Podocarpus totara*. Adults on the wing in September–March. They are nocturnal and attracted to light. **REMARKS** Spindle moths are only found in New Zealand. There are nine described species.

GEOMETRIDAE

Barred Pink Carpet ▪ *Xanthorhoe semifissata* WS 25–32mm e

DESCRIPTION Common endemic moth across Aotearoa New Zealand. Forewings light grey to pale pink with several wavy lines near base. Very distinct brown central band. Female darker than male. Males can be confused with the New Zealand Looper (p. 55), but have a pale forked line along veins in centre of wing. This distinguishes the two species. **DISTRIBUTION** Throughout the lower-mid North Island, South Island and Rakiura/Stewart Island. Most common in the South Island, especially in Otago and Canterbury. **HABITS AND HABITATS** Larvae feed on native and introduced herbs in the cress family (Brassicaceae), including *Cardamine* spp. and Watercress *Nasturtium officinale*. Adult moths inhabit shrubs at edge of native forest. On the wing throughout the year, but more common in warmer months. Attracted to light.

Xanthorhoe occulta ▪ WS 28mm e

DESCRIPTION Peach- or yellow-coloured endemic moth. Forewings pinkish-orange (or peach). Chain of white spots along veins of forewings. Female has reduced wings. **DISTRIBUTION** Widespread across the South Island and Rakiura/Stewart Island. Also present in the lower-central North Island in subalpine habitats. **HABITS AND HABITATS** Larvae feed on herbs. Adults on the wing in September–February and

attracted to light. **REMARKS** The genus *Xanthorhoe* occurs near globally and has more than 100 species. In addition to the species of *Xanthorhoe* profiled here, there are four other uncommonly encountered species in Aotearoa New Zealand: *Xanthorhoe orophyla*, *X. frigida*, *X. lophogramma* and *X. bulbulata*. *X. bulbulata* is classified as Critically Endangered by the Department of Conservation. It is feared to be extinct and has not been sighted since 1991.

◾ GEOMETRIDAE ◾

Five Finger Looper ◾ *Xyridacma alectoraria* WS 42–46mm (e)

DESCRIPTION Endemic moth that occurs in a variety of leaf-like patterns. Hindwings and forewings have more numerous and smaller scallops along termen compared with wings of the similar Tarata Looper (below). **DISTRIBUTION** Widespread in forests of the North, South and Rakiura/Stewart Islands. **HABITS AND HABITATS** May be found year round, but most common in October–February. Attracted to light. Looper larvae feed on leaves of Araliaceae and *Griselinia*, including the Five Finger *Pseudopanax arboreus* and Haumakaroa *Raukawa simplex*.

Tarata Looper ◾ *Xyridacma ustaria* WS 40–46mm (e)

DESCRIPTION Gorgeous endemic looper moth with a variety of leaf-like patterns; also very variable in colour. Some individuals display stripes, others have blotches and some are speckled. Hindwings and forewings have fewer, but larger, scallops (or curves) along termen compared with wings of the similar Five Finger Looper (above). **DISTRIBUTION** Widespread in forests of the North, South and Rakiura/Stewart Islands. **HABITS AND HABITATS** On the wing year round. Attracted to light. Larvae feed on *Pittosporum* species, including the Tarata/Lemonwood *P. eugenioides*, Kōhūhū *P. tenuifolium* and *P. huttonianum*.

▪ GEOMETRIDAE ▪

Large Hebe Looper ▪ *Xyridacma veronicae* WS 36–38mm e

DESCRIPTION Large endemic looper. Often light brown or light orange/yellow with black spots. Generally less brightly coloured than the other two Xyridacma species, but can have striking patterns. Sometimes has dark brown or reddish-brown horizontal stripes across forewings. Sharp angle at apex (outer corner) of hindwings distinctive. **DISTRIBUTION** Widespread in forests of the North, South and Rakiura/Stewart Islands. **HABITS AND HABITATS** On the wing year round but numbers are low in winter. Attracted to light. Larvae feed on *Veronica* species.

Winter Looper ▪ *Zermizinga indocilisaria* WS 26–28mm e

DESCRIPTION Males are active fliers and occur in speckled silver-grey, brown, black and white. Rarely seen female is brachypterous (short winged and flightless), small (10–11mm), and very unusual in appearance. **DISTRIBUTION** Present in drier regions of the eastern South Island (Otago, Canterbury and Marlborough) and lower North Island, as well as Hawkes Bay. **HABITS AND HABITATS** Active throughout the year, but in the highest numbers in late winter and early spring. Neither sex feeds as adults so suited to a winter emergence when few nectar sources are available in its natural habitat. Readily comes to light. This species has been reared by entomologist Brian Patrick from larvae found on small-leaved *Olearia* and *Pimelea* species, *Corokia cotoneaster* and Matagouri *Discaria toumatou*. Other larval hosts plants are thought to include *Ozothamnus* spp. and *Carmichaelia* spp. Introduced hosts include lupin, clovers, briar rose and radiata pine. Larvae are thus highly polyphagous. **REMARKS** Can sometimes be found in high abundance: the most common endemic moth on the wing in August in parts of Central Otago.

GLYPHIPTERIGIDAE & GARCILLARIIDAE

Sedge moths ■ *Glyphipterix* spp. WS 8–20mm (e)

DESCRIPTION Three endemic sedge moths are profiled here: the Stem Borer *G. achlyoessa*, *G. barbata* and *G. triselena*. **DISTRIBUTION** *Glyphipterix* is well spread throughout Aotearoa New Zealand and diverse, with more than 30 endemic species. The Stem Borer occurs throughout the country. *G. barbata* is widespread in the South Island, and *G. triselena* widespread in Otago and Canterbury. **HABITS AND HABITATS** *Glyphipterix* are generally day-flying moths that live in open habitats, grassland, shrubland, forests, around wetlands. Usually found in proximity to hosts, which are often grasses, tussocks or sedges. Larvae mostly borers in seeds, stems or leaves, and a few are leaf miners. When disturbed, *Glyphipterix* often fly a short distance before landing on a grass stem. They also have a habit of pumping their wings up and down when at rest. The Stem Borer inhabits meadows and open grassland. Larvae feed on rushes *Juncus* spp., as well as Cock's Foot *Dactylis glomerata*. *Glyphipterix barbata* larvae feed on the Red Tussock Grass *Chionochloa rubra*. **REMARKS** The sedge moths in the Glyphipterigidae family total more than 430 species worldwide, mostly in the genus *Glyphipterix*, with the largest number in the Australian/New Zealand region.

Stem Borer Glyphipterix achlyoessa

Glyphipterix triselena

Glyphipterix barbata

Echium Leaf Miner ■ *Dialectica scalariella* WS 9–11mm

DESCRIPTION Distinctive small, red-eyed moth in the Gracillariidae family, with bright white colouration on forewing dorsum. **DISTRIBUTION** Native to Europe. Introduced to Australia for biological control of *Echium plantagineum* and later spread to Aotearoa New Zealand. **HABITS AND HABITATS** In New Zealand, larvae feed on flowering plants of the forget-me-not family Boraginaceae. They mine the leaves of their host plant. Adults attracted to light.

Hepialidae

Winter Ghost Moth ■ *Cladoxycanus minos* WS 38-46mm ⓔ

DESCRIPTION Mysterious winter-emergent endemic moth. The only member of the monotypic genus *Cladoxycanus*. Plain ash-grey/brown, sometimes with white markings. The latter are almost always present in males, though variable or absent in females. **DISTRIBUTION** Patchy distribution from Taranaki southwards in the western and central-lower North Island, and throughout much of the South Island. **HABITS AND HABITATS** Larvae strictly associated with mosses in wetlands/bogs, such as *Sphagnum* spp., and have a semi-aquatic lifestyle. They generally live in a secure retreat from which they periodically emerge to feed. Adults primarily on the wing in April–June or July. They fly at night and are attracted to light. **REMARKS** Aotearoa New Zealand has a small number of moth species that emerge as adults over the coldest months of the year. Often these have short-winged and flightless females. Winter emergence may initially sound unfavourable or unusual for a flying insect, but if adult moths can avoid freezing to death, there are several advantages. These include calm/low wind conditions assisting flight and dispersal, and fewer parasites of eggs and young larvae; also, predators of adult moths are less active or in lower numbers.

▪ HEPIALIDAE ▪

Forest Ghost Moth ▪ *Dumbletonius unimaculatus* WS 51–90mm

DESCRIPTION Males of this endemic ghost moth possess deep pink hindwings. Forewings of both sexes variable and usually there is a more complex pattern in males than in females. Hindwings bright purplish-pink in freshly emerged males, paler and more orangey in females, but male's hindwing fades after death and eventually becomes orange or yellow. Distinguished from other Hepialidae in the country by large size and brightly coloured hindwings. Wingspan 51–67mm for males and 74–90mm for females. **DISTRIBUTION** Widespread in the North Island. **HABITS AND HABITATS** Inhabits native forest, especially lowland forest. Adults on the wing in December–April. Larvae thought to feed on fallen leaves on the forest floor, eventually growing to 10cm in length. Pupation occurs in a tunnel beneath the forest floor. Males fly swiftly in the forest canopy and are attracted to light. Females rarely seen and rest in dead hanging tree-fern fronds by day. They lay up to 10,000 eggs randomly across the forest floor. **REMARKS** The species is host to the Vegetable Caterpillar Fungus *Ophiocordyceps robertsii*. This mummifies infected larvae, growing a fruiting body from the caterpillar's head through the soil or leaf litter.

Above: Male, left; female, right

Left: Male, left; female, right

◾ HEPIALIDAE ◾

Pūriri Moth ◾ *Aenetus virescens* WS 150mm (e)

DESCRIPTION The Pūriri Moth or Pepetuna is named after the Pūriri Tree *Vitex lucens*, a common host plant. Endemic to the North Island and the country's largest moth and largest native winged insect. Easily identifiable by large size and vivid green forewings. Adults exhibit sexual dimorphism,

with females being larger (up to 150mm), and males smaller (100mm). Forewings exhibit bright greens, with patterning of brownish-black in females and white pattern in males. Hindwings pinkish. Occasional individuals exhibit blue-green, bright yellow, brick-red and even albino wing colour. **DISTRIBUTION** Throughout the North Island, where they are widespread. **HABITS AND HABITATS** Astonishingly, Pūriri Moths spend the first 5–6 years of their lives as larvae in tree trunks and only the last few days of their lives as moths. Adult moths emerge all year round, but with a peak in September–November. They are active at night, when they mate and lay eggs. Adults do not have functional mouthparts and cannot feed, so are sustained by reserves that were previously consumed by the larvae. Adults commonly attracted to light. Host plants vary widely, and include a range of native forest trees, particularly the Pūriri Tree, Putaputawētā *Carpodetus serratus*, native beech trees, tītoki, maire, kānuka, mānuka, wineberry and houhere. Introduced host plants include eucalyptus, English oak, silver birch, lemon, apple and willow. **REMARKS** Before human colonization, Pūriri Moths may have been an important food source for native birds, reptiles and bats. Today, they are still preyed on by birds, including Kākā and Moreporks, native bats and introduced mammals such as cats and possums.

■ HEPIALIDAE ■

Porina moths ■ *Wiseana* spp. WS 34–66mm (e)

Bog Porina Wiseana umbraculata

DESCRIPTION Porina moths are well known from rural Aotearoa New Zealand, where they are frequent visitors to house lights. There are seven described species, all endemic. Some feed on exotic pastures in New Zealand and can reach high numbers. Adult porina moths can sometimes be distinguished by the markings on their typically fawn or brown forewings. For example, the Bog Porina *W. umbraculata* is the only porina moth with a distinctive white horizontal stripe bordered with dark brown on otherwise fawn forewings. As well as the Bog Porina, the Pasture Porina *W. cervinata* and Summer Porina *W. copularis* are profiled here. The Summer Porina has variable wing patterns and is visually very similar to four other species in the genus. Hence, aside from the Bog Porina, it can be difficult to distinguish some of the other porina species based solely on photographs. **DISTRIBUTION** The Bog and Summer Porinas occur throughout New Zealand; however, the former is uncommon, the latter absent in the northern North Island. The Pasture Porina is widespread throughout New Zealand. **HABITS AND HABITATS** All three species occupy habitats from alpine regions to lowland plains. Adult moths do not feed and only live for a few days. During this time, the female moth flies over grassland and releases hundreds or thousands of eggs. These hatch within a few weeks. Larvae initially feed on the ground surface, but as they grow larger, they burrow beneath the soil and construct silk burrows that can reach a depth of about 30cm. Larvae emerge from these burrows at night to feed. Larvae of all species feed on various grasses and clover. Adults typically on the wing from September to as late as April, but are often in their highest numbers in mid–late spring or early summer. Porina moths are strongly attracted to light and hundreds or thousands can be observed at a single light source on some occasions.

Pasture Porina Wiseana cervinata

Summer Porina Wiseana copularis

LECITHOCERIDAE & NOLIDAE

Sarisophora leucoscia ■ WS 15mm

DESCRIPTION Moth with unusually long antennae. Forewings yellowish with dark streaks along veins and dark patches. Hindwings pale grey. **DISTRIBUTION** Native to Australia. Introduced to Aotearoa New Zealand. Present in upper-central North Island. **HABITS AND HABITATS** Mostly seen in October–April. Many caterpillars of species in this family are densely covered in branched hairs, and feed on dead leaves of plants from families such as myrtle (Myrtaceae), grasses (Poaceae) and protea (Proteaceae).

Gum Leaf Skeletonizer ■ *Uraba lugens* WS 25–50mm

DESCRIPTION Moth native to Australia and accidentally introduced to Aotearoa New Zealand. Broad grey forewings with black and pale grey bands. **DISTRIBUTION** Initially detected in Auckland in 2001. Well established in Auckland by 2007 and spreading. By 2023, it was present throughout the North Island and has also been found in the upper South Island. This emphasizes the speed at which new insect species can establish in a new country if they find the conditions favourable. **HABITS AND HABITATS** Larvae feed on *Eucalyptus* spp. and their close relatives. Eggs laid in parallel lines on both surfaces of *Eucalyptus* leaves. The highly unusual larvae often have a distinctive 'cap' of discarded head capsules, which pile on top of each other after each subsequent moult. This has led to the caterpillars being nicknamed 'mad hatterpillars'.

MICROPTERIGIDAE & MNESARCHAEIDAE

Zealandopterix zonodoxa ■ WS 7–8mm e

DESCRIPTION Beautiful small moth. Endemic and the smallest micropterigid in Aotearoa New Zealand. Forewings predominantly dark brownish-black with stunning purplish-bronze reflections and variable shiny white markings. Hindwings greyish-brown with bronzy-purple reflections. **DISTRIBUTION** Mostly known from the northern North Island, including several offshore islands, such as Great Barrier Island. **HABITS AND HABITATS** Can be abundant but inconspicuous or cryptic. As such, rarely recorded. Inhabits a wide variety of moist indigenous forest types, especially where podocarps are common. Larvae have been found in rotten wood on the forest floor, and extracted from moss and liverworts, but their actual food source is unknown. Day-flying moth but also attracted to UV light at night. Adults have been sighted in large numbers visiting the flowers of the Nikau Palm *Rhopalostylis sapida* and Dwarf Cabbage Tree *Cordyline pumilio*. They are on the wing mostly in November–March. **REMARKS** The Micropterigidae family is a very ancient lineage and unlike most modern-day moths, the adult moths maintain mandibles that they use for chewing pollen and fern spores.

Yellow Dot ■ *Mnesarchaea paracosma* WS 5–6mm e

DESCRIPTION Endemic species of primitive moth. Forewings yellowish-brown, with wedge-shaped whitish streak extending from middle of costa, and some irregular dark patches. **DISTRIBUTION** From about Kaikōura southwards to Bluff in the South Island only. Found at both inland and coastal sites. **HABITS AND HABITATS** Lives in a wide

variety of damp habitats, including tussock grassland, shrubland and damp native beech or podocarp forests, at a range of altitudes from about sea level to 1,200m. Larvae make silk tunnels in periphyton and graze on surrounding plant and fungal material. Adults on the wing in October–February and are day flying, although they are also attracted to light at night. **REMARKS** The Mnesarchaeidae family is Aotearoa New Zealand's only currently recognized endemic moth family.

▪ NOCTUIDAE ▪

Dark Sword Grass ▪ *Agrotis ipsilon* WS 38–48mm

DESCRIPTION Noctuid moth found worldwide. Larvae considered agricultural pests by some because they feed on nearly all varieties of vegetable and many grains. Forewings brown, sometimes red tinged, mixed with pale grey and dark patches. Hindwings whitish-grey or whitish. **DISTRIBUTION** Widespread in Aotearoa New Zealand, and can be very abundant. Occurs in most of the world, but absent from some tropical regions and colder areas. More widespread in northern hemisphere than in southern hemisphere. **HABITS AND HABITATS** Readily comes to light. Larvae feed on a wide variety of weeds and sometimes crops. Adults feed on flower nectar. They are also attracted to deciduous trees and shrubs. **REMARKS** Known to migrate with the changing seasons to avoid extreme temperatures.

Small-eyed Owlet ▪ *Austramathes purpurea* WS 29–42mm e

DESCRIPTION Endemic moth with rich, glossy reddish-brown or purplish forewings. Distinctive black marking in centre of forewings and several scattered whitish scales. Hindwings pale brown with a dark spot in the middle, which is very conspicuous on undersurface. Might be confused with *A. pessota*, but the latter does not have purple hue to forewings. **DISTRIBUTION** Widespread throughout Aotearoa New Zealand. Common in the northern North Island. **HABITS AND HABITATS** Inhabits native forest. Larvae feed on Māhoe *Melicytus ramiflorus* and Narrow-leaved Māhoe *M. lanceolatus*. They pupate among moss or bark. Adults found throughout the year, but most commonly in late autumn, winter and spring.

NOCTUIDAE

Bityla defigurata ■ WS 44mm (e)

DESCRIPTION Distinctive, glossy, plain brown endemic moth. Dark brown head. Faint curved line across forewings. There is one other species in the genus, *B. sericea*, which is very similar to this one, but can be distinguished by pale fringe on forewing termen. **DISTRIBUTION** Widespread in coastal and low-elevation areas. **HABITS AND HABITATS** Larvae feed on *Muehlenbeckia australis* and *M. complexa*. Adults come to light. **REMARKS** Overwintering congregations of adult moths can be found resting under bark on large trees or in wood piles. Moth displays an unusual shunting movement across the ground when disturbed.

Green Garden Looper ■ *Chrysodeixis eriosoma* WS 42mm

DESCRIPTION Head, thorax and forewings have a coppery reddish tinge. One or two large, prominent, whitish-golden markings in centre of forewings. **DISTRIBUTION** Widespread in coastal and low-elevation areas throughout Aotearoa New Zealand, but rare south of Christchurch. Also found in Australia and South-east Asia. **HABITS AND HABITATS** Abundant in agricultural areas, parks, gardens, forest edges, weedy areas, coastal dunes and open habitats. Larvae are polyphagous and have been reported to feed on more than 60 plant species. Adults live for 10–12 days.

◾ NOCTUIDAE ◾

Green Blotched Moth ◾ *Cosmodes elegans* WS 40mm

DESCRIPTION Noctuid moth with distinctive green blotches. Forewings reddish-brown or rust coloured, with three greenish spots edged with silver, the anterior one being hooked. Hindwings reddish. **DISTRIBUTION** Immigrant. Native to Australia but arrives in Aotearoa New Zealand as a migrant or vagrant during summer. Found sporadically throughout the North Island and northern South Island. In Australia, found in New South Wales, Norfolk Island, Queensland, South Australia, Victoria and Western Australia. **HABITS AND HABITATS** Occurs in gardens, weedy areas, forest edges and clearings. Larvae feed on foliage of *Lobelia* and *Verbena* species. They pupate in a cocoon among the leaves of their host plant. **REMARKS** Sharply contrasting forewing pattern is likely to act as disruptive camouflage – a defence against predators – when the moth is resting on or among vegetation.

Scar Bank Gem ◾ *Ctenoplusia limbirena* WS 40–45mm

DESCRIPTION Cosmopolitan moth. Head and thorax clad with grey and black scales. Abdomen pale with dark tufts. Forewings dark and copper tinged. Prominent silvery 'squiggly Y' or 'tooth-shaped' mark in centre of forewings. **DISTRIBUTION** Has been established since 2011 in Aotearoa New Zealand, probably after migrating from Australia. Most common in the North Island but has been recorded as far south as Otago. Found in many areas, including southwestern Europe, Africa, the Canary Islands, Arabia, India, Sri Lanka, southeastern China, Taiwan, Sulawesi, Bali and Timor. **HABITS AND HABITATS** Occurs in gardens, weedy areas, forest edges and coastal dunes. Larvae polyphagous and feed on a wide range of plants. Adults seen mostly in December–May.

■ NOCTUIDAE ■

Orange Peel Moth ■ *Diarsia intermixta* WS 36–37mm

DESCRIPTION Named for distinctive orange colour; however, there are in fact two male colour forms, in beige and orange, and females are dull purplish. Clear, light–dark reniform spot with dark patch on inner edge. **DISTRIBUTION** Native moth widespread on the North, South and Rakiura/Stewart Islands. Also widespread in Australia and islands in the south Pacific. **HABITS AND HABITATS** Found in native forests, gardens, parks and shrubland. Adults on the wing in summer and autumn. Larvae polyphagous and feed on nettles and a wide range of other low-growing plants, including white mustard, turnip, capeweed and ferns.

Māhoe Stripper Moth ■ *Feredayia graminosa* WS 35–40mm

DESCRIPTION Endemic moth. Adults distinctive moss-green in colour, elongated in shape and with strong markings, including bright green orbicular and reniform spots. Male has much enlarged round hindwings. **DISTRIBUTION** Widespread where māhoe occurs, particularly in coastal and forested areas. Less common or largely absent in the inner South Island (for example Central Otago and Mackenzie Country) relative to the rest of the country. **HABITS AND HABITATS** Adults found throughout the year. Attracted to light. Sometimes found resting on mossy tree trunks. Larvae feed on the leaves of the native Māhoe Tree *Melicytus ramiflorus*. Pupation occurs in the soil. **REMARKS** Larvae a known host of the parasitic fly genus *Pales*.

▪ NOCTUIDAE ▪

Cotton Bollworm Moth ▪ *Helicoverpa armigera* WS 30–40mm

Helicoverpa armigera conferta.

DESCRIPTION Strongly polyphagous, cosmopolitan moth. There are two similar subspecies of the Cotton Bollworm Moth, *H. a. armigera* and *H. a. conferta*. In addition, the similar *H. punctigera* migrates here in small numbers. Dark band on hindwing has a couple of pale patches in *H. armigera* that are lacking in *H. punctigera*. **DISTRIBUTION** Immigrant and self-introduced resident. Widespread throughout Aotearoa New Zealand but sporadic in occurrence. Common in much of the world, including southern Europe, North and South America, Africa, Asia and Australia. **HABITS AND HABITATS** Occupies open areas, native and exotic grassland, pasture, arable crops, parks, gardens, coastal dunes and weedy areas. Adults mostly found in November–April. Can fly on warm, sunny days to visit flowers and also at night. Attracted to light. Female lays several hundred eggs in batches on host plant. Larvae feed on a wide range of plants, including crops and garden plants like tomato and lucerne.

Ichneutica agorastis ▪ WS 35mm e

DESCRIPTION Beautiful endemic dark reddish moth. Head, palpi and thorax dark reddish-brown. Abdomen grey, anal tuft light reddish. Lines greyish tinged, edged with dark reddish-brown. Orbicular spots round and reniform spots oblong. Both dark grey in centre, margined with whitish-yellow. **DISTRIBUTION** Widespread in the South Island and Rakiura/Stewart Island. **HABITS AND HABITATS** Adult moths on the wing in January–April. Life history and host species unknown. Found in open habitats, mostly in the subalpine zone. However, in Southland found down to sea level.

NOCTUIDAE

Ichneutica arotis ■ WS 31–46mm (e)

DESCRIPTION Variable endemic moth. Similar to several other 'striped' *Ichneutica* species, especially *I. blenheimensis*, *I. epiastra* (p. 93) and *I. cornuta*. *I. blenheimensis* has blackish forewing fringes, unlike *I. arotis*. *I. epiastra* can be distinguished by obvious row of black dots along outer margin of wing. Male *I. cornuta* has longer pectinations and female lacks dark scaling found on thorax of *I. arotis*. Another distinguishing feature to help distinguish *I. arotis* from other *Ichneutica* is diagonal lines between veins on forewings near termen (visible in the image). **DISTRIBUTION** Throughout the North and South Islands. **HABITS AND HABITATS** Larvae feed on flax *Phormium tenax*. Caterpillars feed at night and rest by day among dead flax leaves. In captivity has been reared on stems of species in the genus *Cortaderia* (pampas grasses). *I. arotis* pupates in a loose cocoon either hidden at base of a stem of flax or on the ground. On the wing in September–April, but most common in September–December.

Ichneutica atristriga ■ WS 35–42mm (e)

DESCRIPTION Rusty-red to pinkish endemic moth. Greyish reniform and orbicular spots. Dark streak at base of forewing extending as far as orbicular spot. **DISTRIBUTION** Throughout the North, South and Rakiura/Stewart Islands. **HABITS AND HABITATS** Larval hosts probably include tussock grasses such as *Poa cita*, *P. colensoi* and *Festuca novae-zelandiae*. Larvae have been reared on species in the genera *Bromus* and *Festuca*. Adults on the wing in November–May (most common in January–March).

NOCTUIDAE

Ichneutica averilla ■ WS 33–41mm

DESCRIPTION Endemic moth. Sometimes confused with the New Zealand Cutworm (p. 97), but forewing colour is more pinkish-brown in *I. averilla*, rather than grey or brown; also antemedian forewing line normally distinct in this species, whereas in *I. averilla* it is absent or faint. Hindwings greyish-brown. **DISTRIBUTION** Found in the North Island at Mount Taranaki, but predominantly widespread throughout the

South and Rakiura/Stewart Islands. **HABITS AND HABITATS** Favours mountainous habitat but can be found down to sea level in southern New Zealand. Adults on the wing in October and April. Larvae have been recorded as feeding on species in the *Plantago* genus, but probably subsist on a variety of herbaceous plants.

Common Snowgrass Owlet ■ *Ichneutica ceraunias* WS 41mm

DESCRIPTION Colourful moth endemic to mountainous country. Head, palpi, antennae, thorax and abdomen brownish-ochreous. Generally light brownish in colour (but darker individuals occur in some areas, such as Southland). Yellowish-white streaks down forewings that may be faint or bold. Pale streaks help identify this species, but it can be mistaken for some forms of *I. dione*. **DISTRIBUTION** Central to the southern North Island, and widespread throughout the South and Rakiura/Stewart Islands. Common

in subalpine and alpine zones throughout its range and occurs down to sea level only on southern coast of Southland. **HABITS AND HABITATS** Hosts of larvae are tussock species in the genera *Chionochloa* and *Festuca*. Adults on the wing in October–February. Some populations have fully winged females, others brachypterous (flightless) females. Adult moths known to fly by day, but mostly nocturnal. They are attracted to light.

▪ NOCTUIDAE ▪

Ichneutica cuneata ▪ WS 34–39mm

DESCRIPTION Endemic, rock-coloured moth with blue-grey forewings, clouded with patches of brown. Hindwings greyish-brown with very broad blackish terminal band. Can be distinguished from similar species by black to dark grey kidney mark on forewings (reniform spot). Underside of hindwing also has discal spot. Individuals found in Tongariro National Park tend to be darker in colour. **DISTRIBUTION** In and around the Tongariro National Park in the North Island and throughout the South Island, but largely inland away from the coast in subalpine and alpine habitats. **HABITS AND HABITATS** Inhabits tussock grassland and shrubland to least 1,640m. Adults on the wing in December–April and can be found flying by day. Life history unknown as are the specific host species of its larvae. Adults have been shown to feed from and may help pollinate *Myosotis macrantha*. They are attracted to light.

Desert Owlet ▪ *Ichneutica disjungens* WS 34–39mm

DESCRIPTION Stunningly patterned and distinctive endemic moth, unlikely to be confused with any other species. **DISTRIBUTION** Central volcanic plateau of the North Island. Throughout inland hill country and mountainous parts of the South Island, but rarely coastal. **HABITS AND HABITATS** Inhabits tussock grassland in alpine and subalpine zones. Hosts of larvae include tussock grasses, possibly encompassing *Poa cita*, *P. colensoi* and *Festuca novae-zelandiae*. On the wing in October–March. Regarded as a fast-flying species and attracted to light.

▪ NOCTUIDAE ▪

Ichneutica epiastra ▪ WS 32–45mm (e)

DESCRIPTION Endemic and variable in both colour and size. Colour of adults can range from pale to pinkish-brown, up to deep brown. Adult can possibly be confused with *I. arotis* (p. 90). Row of black dots on forewing termen of *I. epiastra*, where *I. arotis* has faint dashes, if any markings at all. *I. arotis* also has diagonal lines between veins on forewings near termen and *I. epiastra* does not. **DISTRIBUTION** Throughout the North, South and Rakiura/Stewart Islands. **HABITS AND HABITATS** Prefers open habitats such as wetlands, sand dunes and forest clearings. Eggs laid in summer or autumn and larvae feed during winter and spring. Larval host species are *Austroderia* (genus of five species of tall grasses native to Aotearoa New Zealand, commonly known as toetoe). Adult moths on the wing in October–February and come sporadically to light.

Ichneutica infensa ▪ WS 32–37mm

DESCRIPTION Rusty-red endemic moth. Forewings have a few white marks along costa and on some veins. Relatively distinctive but can possibly be confused with *I. inscripta*, but this species has a more uneven distribution of colours on forewing in comparison to

I. infensa. **DISTRIBUTION** Throughout the South Island, Rakiura/Stewart Island and North Island, at least as far north as Auckland. Most commonly encountered in the South Island. **HABITS AND HABITATS** Inhabits tussock grassland and native forest. Larval host plants are in the genus *Carex* (sedges), including *C. solandri*. Larvae have also been raised on the grass *Bromus catharticus*. Adults on the wing in late October–February (most common in November and December).

NOCTUIDAE

Green-marked Owlet ■ *Ichneutica insignis* WS 36–40mm

DESCRIPTION Highly variable endemic moth. Different individuals encountered on the same night can look vastly different from each other. Can be difficult to separate from *I. skelloni* (p. 102) and less often the Green Carpet Owlet (p. 99), which is usually greener. Pectinations on antennae of male Green-marked Owlet shorter than those on *I. skelloni*, which are more than twice the width (thickness) of the antennal shaft. Forewings of some Green-marked Owlets have distinct whitish suffusion along dorsum, which is considered diagnostic. **DISTRIBUTION** Throughout Aotearoa New Zealand, including Rakiura/Stewart Island. Usually common, but less so in some dry inland areas. **HABITS AND HABITATS** Adults on the wing throughout the year. Larvae recorded as feeding on Red Clover *Trifolium pratense*.

Ichneutica lignana ■ WS 32–40mm

DESCRIPTION Very pale whitish or fawn endemic moth. Quite distinctive in appearance with dark markings on thorax and forewings. **DISTRIBUTION** Widespread on the North, South and Rakiura/Stewart Islands, and Three Kings Islands. **HABITS AND HABITATS**

Lives in a variety of habitats, including coastal areas, tussock grassland, shrubland and native forest, at a range of altitudes from sea level to more than 1,600m. Adults on the wing throughout the year in northern parts of Aotearoa New Zealand, but restricted to October–April in the South Island. Larval hosts are the native grasses/tussocks *Poa cita*, *P. anceps* and *Festuca novae-zelandiae*.

NOCTUIDAE

Ichneutica lithias ■ WS 32–38mm (e)

DESCRIPTION Small endemic moth with stunning contrasting markings in black and white over greyish or reddish-brown base colour. Often has bronze sheen. Reniform and orbicular spots circled in black and white. Zebra-like striped legs. **DISTRIBUTION** Widespread on the South Island, particularly in interior mountains. Also found at Rangipo Desert in the central North Island. **HABITS AND HABITATS** Lives in tussock grassland and shrubland, mainly inland, at a range of altitudes from near sea level to more than 1,400m. Adults on the wing mostly in spring and early summer (October–February). Larval host is the Porcupine Shrub *Melicytus alpinus*, which is common throughout the South Island hill country and mountains.

Alpine Treasure Owlet ■ *Ichneutica maya* WS 37–43mm

DESCRIPTION A treasure to observe in the alpine zones. Distinctively coloured and patterned with purplish and black markings (outlined in white), contrasting against orange (or purplish-orange) base colour. Unlikely to be confused with any other species. **DISTRIBUTION** Found in (or close to) mountains in southern half of the North Island and in the South Island. Widespread along Southern Alps. **HABITS AND HABITATS** Found in subalpine or alpine areas with high rainfall. In Southland can be found down to sea level. Life history in the wild unknown and larval host species unconfirmed. Adults have been seen feeding on *Veronica* blossom. On the wing in November–March and attracted to light.

▪ NOCTUIDAE ▪

Ichneutica moderata ▪ WS 33–43mm (e)

DESCRIPTION Lovely greenish-grey colour with yellow and black touches, but quite nondescript. Can be distinguished from other similar looking *Ichneutica* species, such as plainer examples of *I. cana*, as both male and female have unmarked creamy white fringes on hindwings. One of the most common endemic species of *Ichneutica*. **DISTRIBUTION** From the Bay of Plenty southwards. Throughout the South Island and occurs on the Chatham Islands and Rakiura/Stewart Island. **HABITS AND HABITATS** Inhabits lowland to montane zones. Larvae likely to feed on a variety of low-growing herbaceous plants, including scabweeds *Raoulia* spp., as well as *Crassula manaia* and the English Daisy *Bellis perennis*. Larvae create silk-covered tunnels in roots of their host plants. Pupa enclosed in a loose silken cocoon and shelters among roots of host species. Adult moths on the wing in October–April.

Ichneutica mollis ▪ WS 34–43mm (e)

DESCRIPTION Endemic moth with narrow, pale-coloured forewings. Dark scaling along inner edge of subterminal line distinctive. **DISTRIBUTION** Widespread, but infrequent in occurrence. Found in the North Island from the Coromandel Peninsula southwards, and throughout the South Island and Rakiura/Stewart Island. **HABITS AND HABITATS** Lives in a variety of habitats, including beech forest, podocarp forest and grassland. Life history not well known, but larvae reported to feed on grasses and/or herbs. Adults on the wing in October–March and attracted to light. Has been seen feeding on blossoms.

◾ NOCTUIDAE ◾

Ichneutica morosa ◾ WS 32–40mm (e)

DESCRIPTION Smoky-brown endemic moth. Shades of yellow-brown on forewings suffused with dark greyish-brown. Reniform spot dark greyish with black-edged yellowish border. As well as its standard form, adult moths have a grey colour morph. Can be confused with *I. mustulenta* and *I. lignana* (p. 94), but less whitish than the latter. *I. mustulenta* a deeper red wine colour than *I. morosa*, and underside of hindwings of *I. mustulenta* also a much deeper colour.
DISTRIBUTION Widespread, particularly in the southern North Island and South Island.
HABITS AND HABITATS Found from lowlands to alpine zone. Larvae feed on tussock grasses in the *Poa* genus, including introduced species. Adults on the wing in December–April. Will come to light.

New Zealand Cutworm ◾ *Ichneutica mutans* WS 34–43mm (e)

DESCRIPTION Perhaps one of the most common and widespread endemic moths. Variable in appearance but generally greyish or brownish. **DISTRIBUTION** Widespread all over Aotearoa New Zealand, including in the Three Kings Islands and Rakiura/Stewart Island. Not found in the Chatham Islands. **HABITS AND HABITATS** Adults on the wing year round and attracted to light. Larvae highly polyphagous, feeding on many plants, including grasses, weeds, crops and fruit trees. Pupates on the ground or in moss. May have up to four generations per year but has fewer in colder regions. **REMARKS** The most observed New Zealand moth on the website iNaturalist, with more than 3,000 sightings uploaded to the site. However, there is evidence that the species has declined in some regions.

NOCTUIDAE

Speargrass Moth ■ *Ichneutica nullifera* WS 53–77mm (e)

DESCRIPTION Large, charismatic endemic moth. Adults have little in the way of clear markings or patterns. Forewing can vary in colour from pale fawn to dark grey. Male has wingspan of 53–67mm, female 57–77mm. **DISTRIBUTION** Tongariro National Park, along the Wellington coast and throughout the South Island. **HABITS AND HABITATS** Larval host species are in the genus *Aciphylla* (spear grasses) and as a result the adult moths are often found in habitat such as tussock grassland with an abundance of Aciphylla. Larvae often found in burrows in stem of host plant. Pupa, which is in a loose cocoon, can be found in leaf litter near roots of host plant. Adults on the wing in October–early April, and attracted to light.

Ichneutica omoplaca ■ WS 31–43mm (e)

DESCRIPTION Endemic moth that can be variable in appearance. Some individuals much darker than others. Darker form more common. Diagnostic feature is pale light brown to white colouring between basal streak and costa, which contrasts with ground colour of forewing. **DISTRIBUTION** Widespread from the Bay of Plenty down to Southland. Individuals have also been found in the Auckland Islands. **HABITS AND HABITATS** Lives in a variety of habitats, including beech-forest clearings and tussock grassland. Larval hosts include the Silver Tussock *Poa cita* and Cock's-foot Grass *Dactylis glomerata*, and it has been reared on *Plantago lanceolata*. Adults on the wing mostly in late October–February.

▪ NOCTUIDAE ▪

Black & White Owlet ▪ *Ichneutica paracausta* WS 33–42mm e

DESCRIPTION Beautiful black-and-white addition to the endemic genus *Ichneutica*. In addition to black and white, some individuals contain brown, purplish-brown or reddish-brown colours. However, black streak on forewings, in combination with wing pattern, is highly distinctive. **DISTRIBUTION** Central North Island high country. Widespread in the South Island, generally inland, on hills or mountains. Some coastal populations occur in southern regions. Also on Rakiura/Stewart Island. **HABITS AND HABITATS** Found in tussock grassland, alpine and subalpine shrubland and subalpine forests. Larvae seen feeding on grasses. Adult moths primarily on the wing in October–January. Mid–late spring emerging moth.

Green Carpet Owlet ▪ *Ichneutica plena* WS 31–40mm

DESCRIPTION One of the most beautiful endemic noctuid moths, and also one of the more common. Very variable in appearance, but generally very green. Can be confused with the Green-marked Owlet (p. 94) and *I. insignis*. However, the former is only known from the Auckland District and is a duller olive-green. *I. insignis* is generally less consistently green. **DISTRIBUTION** Throughout the North, South and Rakiura/Stewart Islands. **HABITS AND HABITATS** Occupies a large range of habitats, including towns, cities, shrublands and forests. Larvae polyphagous on herbaceous plants. Adult moths primarily on the wing in late August–May (most of the year).

▪ NOCTUIDAE ▪

Ichneutica propria ▪ WS 30–39mm e

DESCRIPTION Endemic moth that is quite consistent in appearance. May be confused with faded examples of *I. atristriga* (p. 90), but *I. atristriga* lacks distinctive black streak running through middle of forewing. **DISTRIBUTION** Only known from Tongariro National Park and Pureora Forest in the North Island, but widespread in the South Island, especially east of the Southern Alps. **HABITS AND HABITATS** Very common and widespread in subalpine to alpine grassland. Larvae have been reared on tussocks (*Poa cita* and *Festuca novae-zelandiae*) and known to feed on introduced grasses. Adults on the wing in December–May, and attracted to light. Known to visit and may pollinate the Mountain Lacebark *Hoheria lyallii*.

Orange Astelia Wainscot ▪ *Ichneutica purdii* WS 39–55mm

DESCRIPTION Gorgeous ginger moth. Unlikely to be confused with any other species due to its unique bright ginger-orange and pinkish-red colouration. Hindwings blackish-brown. Light grey stripe through centre of thorax. Large moth with wingspan of up to 55mm. **DISTRIBUTION** Found sporadically throughout the main islands of Aotearoa New Zealand, including Rakiura/Stewart Island. More frequent in the south. Largely absent from drier interior of the South Island, for example Central Otago and the Mackenzie Country. **HABITS AND HABITATS** Larvae feed at night on species of *Astelia*. By day they shelter in the interior of the plant. Most common in subalpine and low alpine habitats, for example *Astelia* among tussock grassland just above the natural tree line. Adult moths on the wing in October–March and attracted to light.

NOCTUIDAE

Ichneutica rubescens ■ WS 38–45mm e

DESCRIPTION Endemic with a round or oval mark near centre of forewing (orbicular spot) that encloses small black dot. This is diagnostic and helps distinguish it from the very similar looking *Meterana pascoi*. **DISTRIBUTION** Patchy in occurrence in the central and lower North Island. Widespread throughout the South Island. Also present on Rakiura/Stewart Island and Auckland Islands. **HABITS AND HABITATS** Inhabits tussock grassland and beech forests, as well as podocarp forests. Life history poorly known, but larval hosts likely to be herbaceous plants such as grasses and herbs. Larvae have been reared on *Gunnera prorepens*, a native groundcover species. Adults on the wing primarily in December–April and attracted to light.

Ichneutica scutata ■ WS 30–39mm e

DESCRIPTION Autumn to early winter emerging endemic noctuid. Similar in appearance to the Green-marked Owlet (p. 94) and *I. skelloni* (p. 102), but generally much paler in appearance and basal forewing streak is reddish-brown in comparison to black streak of the other two species. Also, male antennal pectinations much longer in *I. scutata*.

DISTRIBUTION Central to southern North Island as well as Nelson area and eastern parts of the South Island, particularly Otago and Canterbury. **HABITS AND HABITATS** Appears to inhabit lowland grassland as well as coastal sand dunes. Larvae are polyphagous and feed on various herbaceous plants, including grasses, herbs and shrubs. Larvae recorded as feeding, either wild or when reared, on *Plantago* and *Convolvulus* spp., Makaka *Plagianthus divaricatus*, and grasses such as *Poa* spp. Pupates in the soil near its host plants. Adults on the wing in late March–July and attracted to light.

NOCTUIDAE

Common Dotted Wainscot = *Ichneutica semivittata* WS 30–39mm

DESCRIPTION Endemic moth with distinctive pale white-yellow colour and rows of black spots across forewings. Similar to the Dark Underwing Wainscot (opposite), but that species is larger and has much darker hindwings, as well as fewer black spots on forewings than the Common Dotted Wainscot. DISTRIBUTION Widespread on the North and South Islands. Found from the Three King Islands down to Rakiura/Stewart Island. **HABITS AND HABITATS** Occupies a range of habitats, from tussock grassland and shrubland, to forest edges and clearings. Also present

at a range of altitudes, from sea level to the lower alpine zone. Larval host species include rushes such as *Juncus procerus*, sedges like Pukio *Carex secta*, and tussock grasses such as *Poa cita*, *P. colensoi* and *Festuca novae-zelandiae*. Adults on the wing in September–April and attracted to light.

Ichneutica skelloni ■ WS 32–44mm

DESCRIPTION Variable endemic moth. Size, colour and pattern vary in different parts of Aotearoa New Zealand. Can be confused with the very similar Green-marked Owlet (p. 94) and *I. scutata* (p. 101). Male *I. skelloni* distinguishable from the Green-marked Owlet based on having longer pectinations on antennae, more than twice the width (thickness) of the antennal shaft. *I. scutata* generally paler in appearance and usually lacks green markings that some (but not all) *I. skelloni* have. **DISTRIBUTION** Widespread

throughout the South Island but mostly central and eastern areas. In the North Island only confirmed from Wellington region. Also found on Rakiura/Stewart Island. **HABITS AND HABITATS** Found in forest and shrubland habitat. Adults on the wing mostly in April–July and attracted to light. Known host plants for larvae are herbs in the genera *Plantago*, *Senecio* and *Ranunculus* (buttercups) as well as the English Daisy *Bellis perennis*.

NOCTUIDAE

Flax Notcher Moth ■ *Ichneutica steropastis* WS 32–45mm

DESCRIPTION Distinguishable from other *Ichneutica* species based on long, dark longitudinal forewing streaks. Known for distinctive notches caused by larvae feeding on flax leaves; however, these may also be attributable to the larvae of *I. arotis* (p. 90). **DISTRIBUTION** Widespread on the North and South Islands. Found from the Three King Islands down to Rakiura/Stewart Island, as well as in the Chatham Islands. **HABITS AND HABITATS** Occupies a wide range of habitats. Adults on the wing in October–March and come to light. Larvae nocturnal, hiding away in bases of plants and coming out to feed at night. Larvae feed on large-leaved monocots, mainly the Aotearoa New Zealand Flax *Phormium tenax* and Toetoe *Austroderia toetoe*.

Dark Underwing Wainscot ■ *Ichneutica sulcana* WS 35–48mm

DESCRIPTION Streaked and spotted endemic noctuid. Very similar to *I. supersulcana* and cannot easily be distinguished, but *I. supersulcana* is only known from high elevations in the Tararua Ranges and Tongariro National Park, and the two species are not known to occur together. The Dark Underwing Wainscot might also be confused with the Common Dotted Wainscot (opposite), but this is a larger species with darker hindwings and sparser black spots on forewings. **DISTRIBUTION** Throughout the North, South and Rakiura/Stewart Islands from the lowlands to the alpine zone, being recorded up to at least 1,650m. **HABITS AND HABITATS** Lives in native grassland, shrubland, wetlands and native forests. Larval host plants are forest grasses and sedges. Larvae have been reared on the Bush Rice Grass *Microlaena avenacea* and sedges *Carex* spp. They pupate in the soil. Adults mostly on the wing in January–April, but occasionally turn up in other months. They are attracted to light.

▪ NOCTUIDAE ▪

Large Grey Owlet ▪ *Ichneutica ustistriga* WS 36–53mm 🄴

DESCRIPTION Distinctive, large grey endemic moth. Mauve-grey wing colour and unlikely to be confused with other species as patterns on forewings are distinctive. **DISTRIBUTION** From the Three Kings Islands to Stewart Island. Found throughout the North and South Islands. **HABITS AND HABITATS** Occupies a wide variety of habitats, including urban, rural and horticultural areas, orchards, native and exotic grassland, shrubland, tussock grassland and native forest. Larvae eat a variety of herbaceous plants. Recorded food plants include pōhuehue *Muehlenbeckia australis* and *M. complexa*, *Olearia hectorii*, *Plantago lanceolata* and *Juncus effusus*. On the wing throughout the year. Attracted to light and can also be found at rest on tree trunks by day.

Greater Alpine Grey ▪ *Ichneutica virescens* WS 40–49mm

DESCRIPTION Large grey moth of alpine Aotearoa New Zealand. Similar to three other *Ichneutica* species as follows: *I. cuneata* (p. 92) has reniform spots mostly filled in with dark grey (unlike the Greater Alpine Grey). Forewing markings of the Greater Alpine Grey are much stronger than in *I. nobilia*. Distinguished from the smaller *I. panda* by three patches of dark scaling on inner edge of subterminal line which are absent in *I. panda*. **DISTRIBUTION** Central to southern North Island and throughout the South Island. Generally found inland. **HABITS AND HABITATS** Found largely in subalpine and alpine habitats, such as tussock grassland and rocky areas up to high elevations. Adults on the wing in November through to April. Larvae may pupate under rocks. Predominantly nocturnal but sometimes flies by day. Larvae have been reared on a range of plants, including willowherbs *Epilobium* spp., the Feathery Tutu *Coriaria plumosa*, and chain ferns (family Blechnaceae); they may possibly also feed on grasses.

◾ NOCTUIDAE ◾

Sugar Cane Armyworm ◾ *Leucania stenographa* WS 30mm

DESCRIPTION Introduced moth whose pale forewings have a darker central streak and a white spot halfway along. Hindwings buff, with a broken dark line around edge. **DISTRIBUTION** Found in Australia, Aotearoa New Zealand and the Cook Islands. **HABITS AND HABITATS** Larvae feed on grasses (Poaceae), including the Sugar Cane *Saccharum officinarum* overseas.

Meterana alcyone ◾ WS 36–38mm e

DESCRIPTION Tan to dark brown endemic noctuid moth with impressively sharp zigzag markings across forewings. **DISTRIBUTION** Widespread in the North Island and eastern side of the South Island. Also present on Rakiura/Stewart Island. **HABITS AND HABITATS** Known habitats are native forests, shrubland and coastal dunes. Larvae feed on the leaves of the pōhuehue *Muehlenbeckia complexa* and Karaka *Corynocarpus laevigatus*. Adults have been recorded throughout most of the year, but most commonly in August–November. They are attracted to light.

NOCTUIDAE

Pōhuehue Owlet ▪ *Meterana coeleno* WS 34–40mm 🟢

DESCRIPTION Yellow to brownish member of endemic genus *Meterana*. Has a dark infilled reniform spot. Greenish tinge along costa is a feature and helps distinguish this species from *Ichneutica omoplaca* (p. 98). **DISTRIBUTION** Widespread in the North Island and eastern side of the South Island. **HABITS AND HABITATS** Mostly emergent in spring (August–December). Hosts are pōhuehue vines or compact shrubs *Muehlenbeckia complexa* and *M. australis*. **REMARKS** The author was delighted to have this species colonize his property two years after planting a large number of pōhuehue *Muehlenbeckia complexa* in his rock garden. This is the only *Meterana* species recorded at the Central Otago property out of a total of 105 Lepidoptera species recorded over a three-year period.

Lawyer Owlet ▪ *Meterana diatmeta* WS 36–40mm

DESCRIPTION Named after its host plant, the Bush Lawyer *Rubus cissoides*, the moth has a distinctive bright pale green reniform spot on forewing, and greenish suffusion on either side of thorax. **DISTRIBUTION** Widespread, but infrequently encountered, throughout Aotearoa New Zealand, including Rakiura/Stewart Island. **HABITS AND HABITATS** Mostly emergent in spring and early summer (September–January) and will come to light. Host plants are in the genus *Rubus*. Plants in this genus are perennial scrambling vines with compound leaves and reddish prickles on the branches. A prickly plant that can be of great annoyance to walkers and bush users, as it grabs their clothing or scrapes across their skin as they walk past it. However, like most scrambling vines (for example the genus *Muehlenbeckia*), the plant plays an important role in the ecosystem and is crucial for this moth and the beautiful Lawyer Pug (p. 52).

◼ NOCTUIDAE ◼

Kowhai Owlet ◼ *Meterana decorata* WS 34–38mm

DESCRIPTION Intricate, well-patterned moth that is ecologically tied to the kowhai tree. Forewings contain a delicate mixture of brown, black, green and purplish-reds. Whitish reniform spot. Often a white zigzag marking on forewing and a longitudinal black streak. The bright green larvae with white or yellowish stripes can be found on kowhai trees *Sophora* spp. **DISTRIBUTION** Widespread throughout the North and South Islands in habitats with kowhai, including forests, shrubland, and urban and rural areas. **HABITS AND HABITATS** Host plant is species of kowhai. Emergent throughout the year, but lowest numbers in winter. Will come to light.

Larva

Meterana dotata ◼ WS 44–46mm

DESCRIPTION Dark brown or purplish-brown *Meterana* species. Distal half of large reniform spot light brown. Light brown subterminal band only sometimes present. Most similar to M. *ochthistis* (p. 108), but M. *dotata* is a bit larger and reniform is much more strongly marked with whitish, rather than outlined in white. The similar M. *praesignis* (p. 109) has green females (male brown) and both sexes have forewing apex distinctly paler than surrounding areas. **DISTRIBUTION** From the Waikato Region (Mount Te Aroha) to southern North Island, and widespread in the South Island, but mostly inland near the Southern Alps, or in western areas, such as South Westland, West Otago, Western Southland and Fiordland. **HABITS AND HABITATS** Host plants are in the genus *Fuscospora* (beech trees), such as the Mountain Beech *Fuscospora cliffortioides*. Hence the moths' distribution follows beech forest. On the wing in September–April and will come to light.

▪ NOCTUIDAE ▪

Meterana ochthistis ▪ WS 34–36mm

Larva

DESCRIPTION Brownish endemic moth. Most similar to M. *dotata* (p. 107), but M. *ochthistis* is smaller and reniform is proportionally smaller and less strongly coloured in white. Also similar to M. *praesignis* (opposite), but this species usually has green females (males are brown) and both sexes have forewing apex distinctly paler than surrounding areas. **DISTRIBUTION** Widespread throughout the North, South and Rakiura/Stewart Islands. **HABITS AND HABITATS** Unlike many *Meterana* species, which appear to be monophagous to one plant (or genus of plants), the larvae of M. *ochthistis* are polyphagous. They feed on a wide range of trees and shrubs, including *Coprosma propinqua*, Matagouri *Discaria toumatou*, *Helichrysum lanceolatum*, Putaputaweta *Carpodetus serratus*, Porcupine Shrub *Melicytus alpinus* and Pōhuehue *Muehlenbeckia complexa*. On the wing throughout the year, but most common in spring and early summer. Attracted to light.

Meterana levis ▪ WS 32–34mm

DESCRIPTION Greenish endemic moth. Generally light green or whitish-green, and quite distinctive in appearance. Relatively small compared to many of the other *Meterana* species. **DISTRIBUTION** Widespread throughout the North and South Islands, but sparse in occurrence. **HABITS AND HABITATS** Larvae on the endemic trees Lowland Ribbonwood *Plagianthus regius* and Saltmarsh Ribbonwood *P. divaricatus*. On the wing mostly in July–February. Will come to light.

NOCTUIDAE

Meterana praesignis ▪ WS 38–42mm

DESCRIPTION Beautiful *Meterana* species from forested regions. Females often display an attractive mixture of green, brown and white. Males predominantly brown or reddish-brown. The similar *M. dotata* (p. 107) and *M. ochthistis* (opposite) have larger reniform spots. In *M. praesignis* forewing apex is distinctly paler than surrounding areas, unlike in *M. ochthistis* and *M. dotata*. **DISTRIBUTION** Widespread and often locally common in beech forests from the central

North Island southwards. **HABITS AND HABITATS** Generally rare but can be locally abundant, especially around beech forests in southern Aotearoa New Zealand. Mostly on the wing in October–January. Host plant is the Silver Beech *Lophozonia menziesii*. Like many monophagous moths, the distribution of *M. praesignis* follows that of its host plant. Nocturnal and attracted to light.

Mottled Brown Owlet ▪ *Meterana stipata* WS 42–46mm 🟢

DESCRIPTION Endemic moth with sharp-angled, oblong-shaped, whitish orbicular spots. This distinguishes it from all the other *Meterana* species. **DISTRIBUTION** Widespread throughout Aotearoa New Zealand, including Rakiura/Stewart Island and the Chatham Islands. **HABITS AND HABITATS** Larvae feed on *Muehlenbeckia* species. On the wing throughout the year but less commonly encountered in winter. Attracted to light.

NOCTUIDAE

Northern Armyworm ▪ *Mythimna separata* WS 35–50mm

DESCRIPTION Widespread Old-World moth self-introduced to Aotearoa New Zealand. Forewings greyish-yellow with dark grey or reddish-yellow tinge. Orbicular and reniform spots light or yellowish with indistinct edges. Hindwings grey, with dark external margin. **DISTRIBUTION** Widespread in the North Island. Sparse in the South Island but appears to be spreading southwards and increasing in number. Many recent reports around the Nelson area, Blenheim, Christchurch and Dunedin. Also found in China, Japan, South-east Asia, India, eastern Australia and some Pacific islands. **HABITS AND HABITATS** Recorded throughout the year in New Zealand, most commonly in spring and autumn. Larvae feed on a range of grasses, including cultivated plants like Maize *Zea mays*, *Sorghum bicolor*, Common Barley *Hordeum vulgare* and rice *Oryza sativa*. Attracted to light.

Southern Armyworm ▪ *Persectania aversa* WS 40–44mm

DESCRIPTION Pale fawn-coloured endemic moth. Numerous streaks and stripes in white and various shades of brown. In centre of forewing, a distinctive pale notched line, which is useful for identification. Hindwings pale, with broad brownish borders.

DISTRIBUTION Throughout the country including the North, South, Rakiura/Stewart and Chatham Islands. Regarded as common to abundant but less abundant in the northern North Island. **HABITS AND HABITATS** Inhabits open grassland. Larvae feed on grass species including commercial crops. Recorded larval hosts include the Hard Tussock *Festuca novae-zelandiae*, Silver Tussock *Poa cita*, Blue Tussock *P. colensoi*, Marram Grass *Ammophila arenaria* and Bread Wheat *Triticum aestivum*. On the wing mostly in August–April. Adults nocturnal and attracted to light. **REMARKS** Regarded by some as a pest to crops (like many armyworm species). The introduction of a parasitic wasp as a biocontrol agent has reduced its numbers significantly.

NOCTUIDAE

Physetica spp. ■ WS 29–46mm e

DESCRIPTION An endemic genus of silvery moths containing nine species: *P. caerulea*, *P. cucullina*, *P. funerea*, *P. homoscia*, *P. longstaffi*, *P. phricias*, *P. prionistis*, *P. sequens* and *P. temperata*. Most are grey, silver, brown-grey or purplish-grey with various markings. Wingspan range is 29–46mm, depending on species and sex. **DISTRIBUTION** Some species found across Aotearoa New Zealand, such as *P. sequens*, *P. homoscia* and *P. prionistis*. Others, such as *P. funerea*, have more restricted distributions (western and central South Island only). All nine species occur in the South Island. Six species have been confirmed in the North Island. **HABITS AND HABITATS** Most or all species appear to be monophagous (one host plant) or oligophagous (small number of host plants). For example, larvae of *P. homoscia* and *P. temperata* feed on *Ozothamnus* shrubs; *P. phricias* on Matagouri *Discaria tomatou*; *P. cucullina* on *Leucopogon fraseri*; and *P. sequens* on *L. fasciculatus* and *Leptecophylla juniperina*. Larvae and host plants of *P. caerulea*, *P. prionistis* and *P. funerea* unknown. All species attracted to light (*P. prionistis* less so than the others but may be attracted by sugar). Adults generally active during warmer times of the year. Some species emerge in greatest numbers in spring (*P. cucullina*, *P. homoscia*), whereas others, such as *P. longstaffi*, mostly fly in autumn.

Physetica prionistis

Physetica caerulea

Physetica phricias

Physetica cucullina

Physetica longstaffi

NOCTUIDAE

Comma-mark cutworms ■ *Proteuxoa tetronycha* & *P. comma*
WS 29–37mm (e)

DESCRIPTION The comma-mark cutworms are two very similar endemic moths that were long confused as one. Ultimately determined to be independent species by Dr Robert Hoare in 2017. *P. tetronycha* can be distinguished from *P. comma* by paler colour of prothorax as well as browner base colour of forewings. *P. comma* also has more black streaking next to subterminal line. *P. tetronycha* is smaller, with wingspan of 29–33mm v 32–37mm for *P. comma*. **DISTRIBUTION** *P. tetronycha* far more common than *P. comma* and widespread across Aotearoa New Zealand. *P. comma* also found country wide, but rare in upper half of the North Island. Both species present on Rakiura/Stewart Island. **HABITS AND HABITATS** Larvae probably feed on a variety of host species and have been raised on *Acaena* spp., as well as on *Poa cita*; *P. comma* on *Brassica oleracea*. *P. tetronycha* on the wing in September–March and *P. comma* in December–April. Both species attracted to light.

Proteuxoa tetronycha

Proteuxoa comma

Blood-spotted Noctuid ■ *Proteuxoa sanguinipuncta* WS 40mm

DESCRIPTION Distinctive moth with blood-red spots. Rich grey-brown forewings crossed by three thin white wavy lines, bordered by rows of black 'V'-shaped marks. Each forewing bears two black spots ringed with blood-red, making the moth unlike any other in the country. Hindwings tawny-grey. **DISTRIBUTION** Self-introduced and first recorded in Aotearoa New Zealand in 2007. Now common in the North Island. It only colonized the South Island in 2016, but has quickly moved south, being recorded as far south as Otago Peninsula by 2020. Reasonably widespread in Australia; found in Queensland, New South Wales, Victoria, Tasmania, South Australia and southwestern Australia. **HABITS AND HABITATS** Larvae feed on various grasses. Recorded in December–April in New Zealand (peak numbers February–March). Attracted to light.

NOCTUIDAE

Tropical Armyworm Moth ■ *Spodoptera litura* WS 30–38mm

DESCRIPTION Introduced moth that arrived in Aotearoa New Zealand in about 1955. Distinctive appearance, so unlikely to be confused with any endemic moth. DISTRIBUTION Widespread in the North Island and sparse records in the South Island. Most abundant in upper half of the North Island. Native to South-east Asia, Australia and the Pacific. Egg, larva or pupa stages sometimes accidentally transported between regions or countries. HABITS AND HABITATS Occurs in wide variety of habitats, including rural areas, parks, gardens and coastal dunes. More than 110 known host-plant species, belonging to over 40 plant families, making the moth highly polyphagous. Considered a pest of agricultural crops in some countries or regions. In New Zealand larvae feed on a wide range of crop and garden plants.

Slender Burnished Brass ■ *Thysanoplusia orichalcea* WS 36–44mm

DESCRIPTION Self-introduced moth that first appeared in Aotearoa New Zealand in 1984. Forewings pale reddish-brown and extensively covered with a metallic-golden shimmering surface that looks just like burnished brass. Unpatterned hindwings grey-brown, darker at termen. Unlikely to be confused with any other moth species in the country. DISTRIBUTION Widespread in the North Island. Rare in the South Island but present in the Nelson and Marlborough areas. Originated in Indonesia, from where it spread to Europe, South Asia, India, Sri Lanka, Africa, Australia and New Zealand. HABITS AND HABITATS Habitats include parks and gardens, cultivated and rural areas, and weedy places. Most commonly on the wing in February–June. Polyphagous moth whose larvae feed on vegetable crops and various garden plants, including parsley, lucerne, brassicas, sunflower, potato and soybean. Attracted to light.

Oecophoridae

Barea spp. ■ WS 16–22mm

DESCRIPTION Five species of *Barea* native to Australia have accidentally been introduced to Aotearoa New Zealand: *B. exarcha*, *B. confusella*, *B. consignatella*, *B. codrella* and an undescribed species. There are 285 known *Barea* species in Australia, but many are undescribed. Forewings generally have a black and brown or black and white pattern. Hindwings silky fawn. **DISTRIBUTION** *B. exarcha* widespread in both the North and South Islands; *B. confusella* present in upper South Island and throughout the North Island; *B. consignatella* widespread in the North Island, particularly Auckland, Waikato and Northland; *B. codrella* most common in the North Island but has been recorded as far south as Dunedin. **HABITS AND HABITATS** Larvae feed on dead wood, boring into dead trees and living in tunnels under the bark. *Barea* moths have been noted feeding on dead wood of the following plants in New Zealand: Mānuka *Leptospermum* spp., Māhoe *Melicytus ramiflorus* and Gorse *Ulex europaeus*.

Barea exarcha

Undescribed Barea *species*

Ruddy Streak ■ *Tachystola acroxantha* WS 13–15mm

DESCRIPTION Native to Australia and an adventive species in Aotearoa New Zealand and Europe, probably imported with Australian plants. To help identify it, note bright orange colour on termen on forewings. **DISTRIBUTION** Widespread and common on the North, South and Rakiura/Stewart Islands. **HABITS AND HABITATS** Larvae feed on withered leaves and leaf litter, and adults attracted to light.

◾ OECOPHORIDAE ◾

House moths ◾ *Endrosis sarcitrella* & *Hofmannophila pseudospretella* WS <26mm

DESCRIPTION Two common house moths are described here, both introduced species that are widespread globally (cosmopolitan). These are the White-shouldered House Moth *Endrosis sarcitrella* and Brown House Moth *Hofmannophila pseudospretella*. Only the former has white 'shoulders'. Both benefit from humans and their environmental modifications.

Endrosis sarcitrella

DISTRIBUTION Both widespread and common on the North, South and Rakiura/Stewart Islands.

HABITS AND HABITATS House moths occur regularly inside and around buildings such as houses, garages, sheds and farm stores. They breed continuously so can

Hofmannophila pseudospretella

be found at any time of year. Both species attracted to light. Larva spins itself a small silken hideout. Larvae quite adaptable and feed on a wide range of man-made and natural foods and materials, including dried food such as grains, dried fruits, cereals (including bran and flour), seeds, potatoes, rotting wood, fur, wool, and even insect specimens, bird's nests and guano. Other items reportedly eaten include book bindings, wine-bottle corks, furniture fabrics and leather. **REMARKS** Both species are successful hitchhikers, piggybacking with humans around the world. In fact, both species have spread so widely that their 'true home', or pre-human distribution in the wild, is unknown. However, *Endrosis sarcitrella* is likely to have originated in southern South America, where another species of the genus was recently described.

▪ OECOPHORIDAE ▪

Tingena spp. ▪ WS 16–22mm

DESCRIPTION Genus endemic to Aotearoa New Zealand. There are more than 90 species, making *Tingena* the largest genus of Lepidoptera in the country. Moths in this genus are small but have a characteristic wing shape. Some species look quite similar to each other. Genus requires more study, and there are a number of undescribed species. Due to this, identification to species level is not always possible, even with clear photographs, expert opinion and location data. Some species brightly coloured in orange or yellow. Four species are profiled here: *T. actinias*, *T. chloradelpha*, *T. compsogramma* and *T. phegophylla*.
DISTRIBUTION The *Tingena* genus is widespread across New Zealand. **HABITS AND HABITATS** Varied habitats, from domestic gardens and cultivated land, to mature native shrubland and forest. Larvae of all species feed on leaf litter. Many *Tingena* moths will come to light at night. Some also actively fly by day. Pōhuehue *Muehlenbeckia australis* is a known host, with the larvae feeding on the fallen leaves. Adults on the wing in warmer months of the year, in September–March.

Tingena actinias

Tingena compsogramma

Tingena chloradelpha

Tingena phegophylla

■ OECOPHORIDAE ■

Trachypepla spp. ■ WS 13–15mm

Trachypepla contritella

Trachypepla conspicuella

Trachypepla protochlora

Trachypepla lichenodes

DESCRIPTION About 23 species of *Trachypepla* moth are endemic to Aotearoa New Zealand, but some may be misplaced in the genus. Four species are profiled here, which vary in colour and pattern. Many *Trachypepla* species have distinctive tufts of raised scales. *T. contritella* variable in strength of ground colour, forewing markings and paler central area on forewings; *T. conspicuella* also variable, but can be recognized by brown thorax and pale basal patch. *T. contritella* thought to mimic grey lichens, *T. conspicuella* possibly bird droppings. *T. protochlora* a beautiful greenish colour, and *T. lichenodes* has yellow and black patches. **DISTRIBUTION** All four species widespread throughout the North and South Islands, but *T. protochlora* and *T. lichenodes* rather rarely seen. Since the 2010s, *T. contritella* has also been seen in the United Kingdom. **HABITS AND HABITATS** Preferred habitat of all species is native forest, such as beech forest, and shrubland. Larvae of *T. contritella* have been reared from pendulous epiphytic mosses, *Weymouthia* or *Papillaria* spp. Adults of all four species on the wing in about October–February and attracted to light. *T. conspicuella* often seen resting on man-made structures such as fences or walls, and known to enter houses.

■ Papilionoidea ■

Long-tailed Blue ■ *Lampides boeticus* WS 24–34mm

DESCRIPTION Small butterfly whose common name refers to long streamers on its hindwings, male's bright blue colour and typical host plant of the butterfly (members of the pea family Fabaceae). Exhibits sexual dimorphism, with males having a primarily blue upperwing surface with brown edges, while females have a smaller amount of blue. Both sexes have a thin, long tail on each hindwing, which is very useful for distinguishing the butterfly from other species. Also two black eye-spots on underside surrounded by orange. **DISTRIBUTION** Most common in the upper half of the North Island, but present in coastal locations throughout the North Island and upper or western South Island (Nelson, Marlborough and Westland). First recorded in Aotearoa New Zealand on Waiheke Island near Auckland in 1965 and was well established by 1970. Likely to have migrated from Australia. Also found in Europe, Africa, South and South-east Asia, Australia and islands throughout the Pacific Ocean. **HABITS AND HABITATS** Inhabits forests edges, coastal dunes, meadows, grassland, gardens and other places with flowering plants. Larvae feed on flowers, seeds and pods of many legumes (Fabaceae), especially Gorse *Ulex europaeus* in New Zealand. May be seen year round and likely to have multiple broods in warm areas. Most commonly seen in early autumn. Flies by day in sunny, warm weather with a jerky flight as it visits flowers. **REMARKS** Tails on hindwings thought to imitate antennae and together with eye-spots, confuse predators into thinking the back end of the butterfly is the front end. This deflects some attacks by birds (and other predators) to a non-essential part of the body, increasing the butterfly's chance of survival.

Female

■ PAPILIONOIDEA ■

Southern & Common Blues ■ *Zizina oxleyi*  & *Z. otis* WS 17–27mm

Southern Blue Zizina oxleyi

DESCRIPTION Two similar blue butterflies, one endemic the other introduced. The Southern Blue (or New Zealand Blue) was once widespread throughout the country, but is thought to have been displaced from much of it by the introduced Common Blue (or Lesser Grass Blue), which arrived from Australia in the early days of European settlement. Both are small butterflies with pale violet-blue forewings with a silvery sheen. The Southern Blue has more rounded forewings then the Common Blue. The Southern Blue also has darker zigzag markings on underside of wings with a distinctive banded border. **DISTRIBUTION** Southern Blue mostly found from North Canterbury to Central Otago. Occasional remnant populations still present in the North Island but rare. Common Blue common throughout the North Island and upper South Island. Also occurs in Australia and numerous islands throughout the Pacific. **HABITS AND HABITATS** Both butterflies can be found flying close to the ground over lawns, grassland, roadsides, stony areas and riverbeds up to 900m. They prefer areas that have varied habitat including shelter, nectar plants and stones for sunbathing. For the Southern Blue, native food plants include low-growing brooms *Carmichaelia* spp. and the Scree Pea *Montigena novae-zelandiae*. Like the Common Blue, it also oviposits on introduced legumes of the Fabaceae family, such as clover *Trifolium*, *Medicago* and *Lotus* spp., and the Tree Lucerne *Chamaecytisus palmensis*. **REMARKS** The Central Otago population of the Southern Blue is at the southern limit of its distribution, so with southern spread of the Common Blue this population is of great conservation significance. By an unknown mechanism, the Common Blue seems to eliminate the Southern Blue but fortunately, for now at least, it appears to have stalled its southern spread in North Canterbury.

Common Blue Zizina otis

▪ Papilionoidea ▪

Copper butterflies ▪ *Lycaena* spp. WS 17–35mm

DESCRIPTION Delightful group of brightly coloured butterflies found throughout Aotearoa New Zealand. Primarily orange and/or purple-blue, with black outlines, patches and lines. The genus *Lycaena* is a very variable group of butterflies undergoing rapid evolution. It needs more taxonomic and genetic work to identify how many species there are within it. The group has been difficult to resolve and there are many different forms, differing in details of size, wing shape, colour pattern and ecology. Twenty-five forms were recognized in the 2012 *Butterflies of the South Pacific* by Brian and Hamish Patrick, some or all of which may be good species. However, if so, currently most of them are undescribed and there are only eight scientifically named species. **DISTRIBUTION** Spread throughout the entirety of the North and South Islands in suitable habitats. Some species have wide distributions, whereas others are rare, such as the undescribed Chrystalls Beach Boulder Copper. **HABITS AND HABITATS** Some species occupy coastal habitats, like the North Island Coastal Copper *L. salustius*. Others occupy inland mountainous terrain, including alpine zones, tussock grassland and braided rivers, such as the Canterbury Alpine Boulder Copper *L. tama*. Larvae feed mostly or entirely on one or more species of *Muehlenbeckia* (or pōhuehue), particularly *Muehlenbeckia australis*, *M. complexa* and *M. axillaris*. All *Lycaena* butterflies can be spotted in sunny weather flying low to the ground over their habitat, basking on stones or visiting flowers on prostrate vegetation such as mat daisies *Raoulia* spp.

Common Copper Lycaena *sp.*, Nelson

Central Otago Boulder Copper Lycaena *sp.*

Central Otago Boulder Copper Lycaena *sp.*

Glade Copper Lycaena feredayi

▪ Papilionoidea ▪

Tussock ringlets ▪ *Argyrophenga* spp. WS 35–45mm ⓔ

DESCRIPTION Genus of exquisitely-marked endemic butterflies. Comprises three species that all fly in the mountains of the Southern Alps. The Tussock Ringlet or Common Tussock *Argyrophenga antipodum* is the most widespread species and is described here. It is brown and orange, with distinctive eye-spots – this is believed to be an evolutionary adaptation either to scare predators or to deflect them from attacking the body. Adorned with distinct silver streaks on undersides of wings. Male and female somewhat different in body shape and colour. Female shorter and lighter (more yellowish) than male, with more rounded body. Male darker (with more red and brown) but with similar markings. **DISTRIBUTION** Eastern, southern and central regions of the South Island, most abundantly in hills and mountains of Canterbury, Otago and Southland. **HABITS AND HABITATS** Found in a variety of habitats, ranging from sea level to the alpine zone (to 1,950m). Habitats include salt marshes, wetlands and tussock grassland. A new generation is hatched annually, with adult butterflies on the wing in late October–late March. In larval stage, relies on members of the family Poaceae (grasses, including tussocks) as host plants. **REMARKS** There are two other species in the genus, but each has a more limited distribution. Janita's Tussock *A. janitae* is also widespread in the South Island, but absent from the west coast, western Otago, western Southland and Fiordland. It can be distinguished from the Common Tussock by having no white outer marginal line on underside of hindwing. Harris's Tussock *A. harrisi* is only found in northern parts of the South Island. It favours wetter areas and only overlaps with Janita's Tussock from Tasman Mountains to Lewis Pass area.

The Common Tussock Argyrophenga antipodum *– note white outer marginal line on underside of hindwing.*

Janita's Tussock Argyrophenga janitae *– note absence of white outer marginal line.*

■ Papilionoidea ■

Monarch Butterfly ■ *Danaus plexippus* WS 89–102mm

DESCRIPTION One of the most well recognized and studied insects in the world, known equally for its beauty and amazing ecology. Capable of extensive migratory travels, it has made itself a home in Aotearoa New Zealand since first arriving in around 1870. Wings feature easily recognizable black, orange and white pattern, with a wingspan of 89–102mm. **DISTRIBUTION** Throughout the North Island and most of the South Island. Can be common near human habitation. Sparse in the lower South Island. **HABITS AND HABITATS** Occupies parks and gardens in many towns and cities throughout New Zealand, where host plants are present, but will flutter into other habitats. Adult butterflies overwinter in sheltered places and sometimes aggregate in sheltered spots in tall trees. Can be seen flying around on sunny, mild winter days. Slow, gliding flight but capable of migrating huge distances. In New Zealand, main host plants are species of swan plant *Gomphocarpus*. In other areas, such as North America, larvae feed on milkweeds *Asclepias*. Eggs laid on undersurfaces of leaves; larvae mature over about three weeks, developing characteristic black, white and yellow-banded pattern as they mature, then pupate suspended by the rear end in a sheltered spot. Adult emerges from its chrysalis after about two weeks of pupation. **REMARKS** Eastern North American Monarch population is extraordinary for its annual southwards late summer/autumn migration from the northern and central United States and southern Canada to Florida and Mexico. During autumn migration, monarchs cover thousands of miles, with a subsequent multigenerational return north in spring.

PAPILIONOIDEA

Forest Ringlet ■ *Dodonidia helmsii* WS 45mm e

DESCRIPTION Perhaps the most beautiful endemic butterfly, a special creature to behold while tramping in the bush. Also known as Helms' Butterfly, it is the only member of the genus *Dodonidia*. Wing uppersides red-brown but with bold dark brown and yellowish bands running parallel to outer margins; underside banded in silvery-white. Series of distinctive eye-spots, with four on each hindwing and one on each forewing. **DISTRIBUTION** Scattered throughout the North Island and upper South Island as far south as Lewis Pass. Generally rare. **HABITS AND HABITATS** Occupies native forest, especially beech forest. Adults usually found in October–March, but often their flight period is restricted to just one or two months, with the month/s varying with geographical location. Often flies high off the ground in the forest canopy, so can be very difficult to see or photograph. May descend to feed on flowers or bask in the sun in a forest clearing. Females lay eggs on 'cutty grass' *Gahnia* spp. and snow tussock *Chionochloa* spp. Larvae feed in autumn, winter (possibly only in northern populations) and spring on the leaves. They leave notches as they go. Adults live for 3–4 weeks and are fast flying. Species is of some conservation concern, as sharp declines have been noted in areas such as the Waitakere Ranges west of Auckland. Populations at higher elevations, above 600m, appear to be faring better than those at lowland sites. Predation on larvae by the introduced European Wasp *Vespula germanica* may be a cause of decline, in combination with habitat loss.

Larva

PAPILIONOIDEA

Blue Moon ▪ *Hypolimnas bolina* WS 70–90mm

Female

DESCRIPTION Large, black-bodied butterfly with vibrant markings. Has a high degree of sexual dimorphism. Female has multiple morphs, while male is relatively consistent in appearance. In males dorsal wing surface jet-black but features three prominent spots, two on forewing and one on hindwing. These are white fringed with blue-violet. Males also have numerous smaller white spots on fringes of forewings and hindwings. Ventral surface consists essentially of banded white markings set against a brownish background. Females variable in upperside pattern, with different morphs varying in presence of white, orange and blue markings. Female underside similar to that of male. **DISTRIBUTION** Sporadic migrant in warmer months from tropical Australia, where it is known as the Common Eggfly. So far, no confirmed breeding in Aotearoa New Zealand. Widespread globally in the tropics and subtropics from Madagascar and Africa, through India, to China and the Pacific. **HABITS AND HABITATS** Larvae feed on wide range of low-growing plants in the families Acanthaceae, Amaranthaceae, Malvaceae and daisies (Asteraceae). Adult butterflies feed on nectar from a variety of flowering plants.

Australian Painted Lady ▪ *Vanessa kershawi* WS 50–55mm

DESCRIPTION Attractive migrant butterfly from Australia, which can turn up almost anywhere in Aotearoa New Zealand but is not known to be a permanent resident. Wings black, red and brownish-orange, with white markings near forewing-tips. Four eye-spots near termen on each hindwing. **DISTRIBUTION** Mostly confined to Australia and New Caledonia, although strong winds bring it to New Zealand in varying numbers each summer. Although there are reports of breeding in New Zealand, does not appear to overwinter successfully, and thus has not been able to establish permanent self-supporting colonies. **HABITS AND HABITATS** Inhabits open country, including parks and gardens. Larvae feed mainly on everlasting daisies in the genera *Gnaphalium* and *Helichrysum*. Fast-flying migratory species. To find mates, males exhibit territorial behaviour, which involves a male perching on vegetation in a sunny spot on a hilltop, waiting for females to fly by.

Black Mountain Ringlet ■ *Percnodaimon* sp. WS 35–55mm

DESCRIPTION A pleasure to witness gliding over an alpine scree slope on a bright summer's morning, this butterfly has broad, velvety wings that are usually black or dark brown across the dorsal surface, sometimes with a purplish iridescence. Forewings have distinctive black and white circles inside patch of brown. Males and females similar in appearance. Although only two species, *P. merula* and *P. micans*, have been scientifically described to date, research suggests that *Percnodaimon* may be a complex of related species. In their 2012 guide to butterflies of the South Pacific, Brian and Hamish Patrick suggested there may be up to eight species of *Percnodaimon* still to be described, based on wing shape, pattern, colour, size, behaviour and season of emergence. More work is required on these butterflies. **DISTRIBUTION** *Percnodaimon* butterflies are spread throughout the Southern Alps and other high mountainous country throughout the South Island, from Nelson and Marlborough to Southland. **HABITS AND HABITATS** Lives almost exclusively in rocky areas, usually above 1,200m, but has been seen at up to 3,100m. Can be quite common and seasonally abundant in good habitat, such

as large greywacke scree slopes with adjacent tussock grassland. Eggs laid on rocks and larvae feed on alpine tussocks *Poa* spp. Larva pupates under a stone. Black wing colour, with a large surface area, helps absorb sunlight and store heat, a clever adaptation to the cold climate of the Southern Alps.

▪ Papilionoidea ▪

Butler's Ringlet ▪ *Erebiola butleri* WS 35–43mm 🟢

DESCRIPTION Quaint endemic butterfly of remote alpine localities. Males a dark smoky-brown colour. Females paler and from above can look similar to a faded tussock ringlet, *Argyrophenga*. Key for identification is underside of hindwings, which has scattered, wedge-shaped, silvery-white marks (unlike *Argyrophenga* species, which have long, thin silvery streaks). The Black Mountain Ringlet, *Percnodaimon* species, is similar, but blacker, and lacks underside wedge-shaped marks. The only member of the genus *Erebiola*. **DISTRIBUTION** Subalpine and alpine sites, zigzagging along the main divide of the South Island. Scattered from Paparoa Range down to northern Fiordland and western Otago. **HABITS AND HABITATS** Favours damp slopes or terraces at about 900–1,800m, with snow tussock *Chionochloa*, *Veronica* (Hebe) and *Dracophyllum* shrubs. Generally sighted in January–March. Host plant for larvae is snow tussock *Chionochloa* spp. Rarely seen, probably due to a late-summer emergence, and remoteness of many of its habitats.

Honshu White Admiral ▪ *Limenitis glorifica* WS 60mm

DESCRIPTION Alluring butterfly deliberately introduced in the fight against the invasive Japanese Honeysuckle *Lonicera japonica*. Butterfly is endemic to the island of Honshu, in Japan. **DISTRIBUTION** First released in the Waikato and Auckland in 2014. Now released in various locations in the North and South Islands by regional councils and is likely to be gradually establishing in many regions. **HABITS AND HABITATS** Eggs laid almost exclusively on Japanese Honeysuckle. Its release into Aotearoa New Zealand was approved in August 2013. The plant is listed in New Zealand as an unwanted organism. Once it has invaded an area, it grows rapidly and outcompetes native plants for sunlight and nutrients. It is hoped that the Honshu White Admiral will help control this invasive threat to forests and shrubland.

■ Papilionoidea ■

New Zealand Red Admiral ■ *Vanessa gonerilla gonerilla* WS 50–60mm (e)

DESCRIPTION Iconic endemic butterfly. The Māori name 'kahukura' means red cloak. Medium sized with a 50–60mm wingspan. Top side of forewings mostly black, with a bright red bar and white markings near wing-tips. Rear wings dark reddish-brown with red patch containing four black circles with pale blue in centre of each. Underside pattern camouflages butterfly when resting on vegetation. Two subspecies: *V. gonerilla gonerilla*, which occurs on mainland Aotearoa New Zealand, and *V. g. ida*, on the Chatham Islands. **DISTRIBUTION** Common throughout the country where its food plants occur, but has become rare in some areas. **HABITS AND HABITATS** Primarily inhabits native forest (where it is often seen among clearings and edges), shrubland, parks and gardens. Primary larval host plant is the native stinging nettle, Ongaonga *Urtica ferox*, although larvae can also eat other *Urtica* species. Adults occur most frequently in summer but can overwinter too. Day flying. **REMARKS** Thought to be in decline due to spraying and decline of native nettle plants. Nettle species are now being cultivated by various conservation groups to help combat this decline. Vulnerable to exotic parasitic wasps *Echthromorpha intricatoria* and *Pteromalus puparum*, which may have contributed to its decline.

▪ PAPILIONOIDEA ▪

Yellow Admiral ▪ *Vanessa itea* WS 48–55mm

DESCRIPTION Medium-sized butterfly native to Australia, Aotearoa New Zealand (self-introduced resident), Lord Howe Island and Norfolk Island. The Māori name 'kahukōwhai' means yellow cloak. **DISTRIBUTION** Common throughout New Zealand where its food plants occur. **HABITS AND HABITATS** Strong flier thought to survive wind-blown voyages from Australia. Flies by day in open habitats, grassland, weedy areas, roadsides and gardens. Nettles a preferred food for the larvae, which feed at night and hide in a curled leaf by day. In New Zealand they feed mostly on the introduced Dwarf Nettle *Urtica urens*. Butterflies on the wing in warmer months of the year and may live for several months. Adults feed on nectar from various flowers and sometimes sap from trees. **REMARKS** Has benefited from clearance of native forest in New Zealand and creation of open habitats. While it probably migrated from Australia long before human arrival, the removal of native forest is thought to have assisted it in forming more permanent colonies.

Cabbage Butterfly ▪ *Pieris rapae* WS 32–47mm

DESCRIPTION Familiar sight to all New Zealanders, whether welcome in the garden or not. Accidentally introduced and native to Europe. The only bright white butterfly in Aotearoa New Zealand. In the family Pieridae. **DISTRIBUTION** Very abundant throughout. Only largely absent from heavily forested areas and high alpine area. **HABITS AND HABITATS** Day flying in gardens, parks, roadsides, farmland and cropland. Adults seen in September–April. They have multiple generations each summer. Eggs laid on members of the cabbage family, including cabbage, turnip, cauliflower and radish. Also on nasturtium *Tropaeolum* spp. Larvae are green and feed on the leaves. **REMARKS** Accidentally introduced in about 1930. Spread very quickly and thrived, becoming a major pest of crops. Two small parasitic wasps were soon introduced to New Zealand to help control its numbers and appear to prevent extreme outbreaks.

PLUTELLIDAE

Leuroperna sera ◾ WS 10mm

DESCRIPTION
Small, self-introduced resident moth from Australia. Similar to the Diamondback Moth (below), but has much broader wings and lacks pale yellow pattern on forewing dorsum. When resting, adult moths hold their antennae in a forwards 'V'-shaped position.

DISTRIBUTION In Aotearoa New Zealand occurs in the North Island and South Island down to Canterbury. Also found in Japan, Taiwan, Vietnam, Indonesia, India, Sri Lanka and Australia. **HABITS AND HABITATS** Occurs in varied habitats, from edges or clearings of native bush, to domestic gardens. Larvae known to feed on members of the cabbage family, Brassicaceae. Adults have been seen swarming over watercress so this may be another larval host plant. On the wing all year round, but most often seen in warmer months of the year. Attracted to light.

Diamondback Moth ◾ *Plutella xylostella* WS 15mm

DESCRIPTION Small, greyish-brown moth. Sometimes has cream-coloured band that forms one or two diamond shapes along back. Wingspan about 15mm and body length 6mm. Geographical origin unknown, but has spread virtually worldwide. **DISTRIBUTION** Widespread in the North and South Islands, and many offshore islands including Rakiura/Stewart Island and Chatham Islands. Occurs almost globally. **HABITS AND HABITATS** Lays eggs only on plants in the Brassicaceae family, including many vegetable crops such as broccoli, brussels sprouts, cabbage, cauliflower, kale, mustard, radish and watercress. Wild species in the Brassicaceae family also act as hosts when cultivated crops are unavailable. Adult moths largely nocturnal and attracted to light. **REMARKS** Considered one of the worst pests of brassicas in the world, and much research has gone into control methods, including pesticides and biological controls. Larvae damage leaves, buds and flowers, and may disrupt head formation in cabbage, broccoli and cauliflower.

PSYCHIDAE

Australian Lichen Bag Moth ■ *Cebysa leucotelus* WS 8–16mm

Female

DESCRIPTION Stunning brightly coloured bag moth native to eastern and southern Australia. Adult female metallic looking, having bright blue wings with yellow wing-tips and patches. Wings do not expand properly so she is unable to fly. Male has similar pattern but is purplish-brown and does not appear metallic (lacks iridescence). Males have fully developed wings and fly normally. **DISTRIBUTION** Australian moth first found in Auckland in 1981 (probably accidentally imported). Now found throughout the North Island but most common in Waikato, Auckland and Northland. **HABITS AND HABITATS** Occupies dry microclimates in cities, parks and gardens. Larvae build a protective bag like other bag moths, and feed on lichens, fungi and algae, retreating into the bag when disturbed. Adults hatch late summer or autumn and are usually on the wing in March–April. Females often found on the ground. Males fly actively in sunny weather in search of females. Neither sex feeds and adults only live long enough to mate and produce eggs.

Lepidoscia protorna & *L. heliochares* ■ WS 20mm

DESCRIPTION *L. protorna* has speckled grey or brown forewings, each with dark spots. *L. heliochares* speckled grey, with vague markings, including paler grey band on each forewing. **DISTRIBUTION** Both Australian moths first found in Aotearoa New Zealand in the mid–late 1970s. *L. protorna* occurs throughout the North Island; *L. heliochares* has spread throughout most of the country as far south as Dunedin. **HABITS AND HABITATS** Larval case of *L. heliochares* known as a 'thatched cottage' because of its untidy structure of fragments of plant material. Larger, neater case of *L. protorna* known as a 'big log cabin'. Both feed on leaf litter and/or lichens/algae. *L. protorna* on the wing in December–May. *L. heliochares* on the wing in June–September. Both species nocturnal and attracted to light. **REMARKS** Genus poorly understood taxonomically and in need of revision.

Lepidoscia heliochares

Lepidoscia protorna

PSYCHIDAE

Common Bag Moth ■ *Liothula omnivora* WS 28–38mm

DESCRIPTION Endemic moth that is rarely seen, despite its pupal cases being a regular sight throughout the country. Male a hairy, dark brown moth with semi-translucent wings, tapered abdomen and 28–38mm wingspan. Only male metamorphoses into a recognizable moth. Adult female never leaves the bag and has no wings. **DISTRIBUTION** Widespread throughout the North, South and Rakiura/Stewart Islands. **HABITS AND HABITATS** Like other bag worms, larvae construct and live in a small, mobile bag of silk, which provides camouflage and protection for the larva. When threatened, a larva retreats into the bag and closes it firmly. Bag is constructed within a few days of hatching. Outer surfaces of bags often decorated with grass, small leaves or stick fragments. Larvae feed at night on a wide range of native and exotic shrubs and trees. Pupation usually occurs during winter. When ready to pupate, a larva secures the bag to a twig using silk. It then pupates within the bag. Only males emerge from the bag and fly to find a female.

Bag

PTEROPHORIDAE ENDEMIC PLUME MOTHS

Hebe Plume Moth ■ *Amblyptilia falcatalis*, & *A. repletalis*
WS 22mm & 14–21mm e

Amblyptilia falcatalis

Amblyptilia repletalis

DESCRIPTION *A. repletalis* smaller than the Hebe Plume Moth *A. falcatalis*, with shorter labial palps and less distinct dark triangle near apex of forewing. **DISTRIBUTION** Both species widespread in the North and South Islands. **HABITS AND HABITATS** Larval host plants of *A. falcatalis* are in the *Veronica* (*Hebe*) genus. It probably has two broods per year. Adults have been seen on the wing all year round. *A. repletalis* found in a variety of habitats, including native forest clearings, shrubland, coastal dunes and gardens. Larvae feed on seed heads of *Plantago* species. Adults on the wing all year round. Both species attracted to light. **REMARKS** There are at least 22 species of plume moth in Aotearoa New Zealand.

Araliad Plume ■ *Pterophorus monospilalis*, & *P. innotatalis*
WS 21–23mm & 15–16mm e

DESCRIPTION Two endemic plume moths, the Araliad Plume silvery-whitish and *P. innotatalis* pale brown. **DISTRIBUTION** Both species widespread in the North and South Islands. The Araliad Plume is also found on Rakiura/Stewart Island. **HABITS AND HABITATS** The Ariad Plume inhabits native forest, parks and domestic gardens. Larvae active by day and feed exposed on Araliaceae species (ivy family like the Common Ivy, Five Finger, Lancewood and Patē). They have several broods per year. Adult moths on the wing in October–June and attracted to light. Adult *P. innotatalis* on the wing in August–May and prefers to inhabit grass or fern-covered hill slopes to 1,500m. Larvae start their lives as leaf miners and later feed on exposed leaves of *Dichondra* species, including the Kidney Weed *Dichondra repens*.

Ariad Plume Pterophorus monospilalis

Pterophorus innotatalis

◾ PYRALIDAE ◾

Senecio Blue Stem Borer ◾ *Patagoniodes farinaria* WS 30mm

DESCRIPTION Moth with distinctive long, blueish-grey forewings suffused with white and dark speckles and some reddish scaling near forewing base. Two ragged dark bands across forewings. Hindwings plain pale grey. **DISTRIBUTION** Native or self-introduced resident shared with Australia. Widespread in Aotearoa New Zealand. **HABITS AND HABITATS** Occupies a wide variety of habitats, including coastal areas, farmland, shrubland and native forest edges and clearings. Adults most commonly seen in October–April. They fly at night and are attracted to light. Larvae are shoot borers that feed on Ragwort *Senecio jacobaea* and native *Senecio* species. Larvae bore into the stems. **REMARKS** Although this moth feeds on Ragwort it does not affect the survival or reproduction of the plant, as the larvae do not feed on the seed heads or flowers.

Meal Moth ◾ *Pyralis farinalis* WS 18–30mm

DESCRIPTION Cosmopolitan moth with colourful banded forewings. Well adapted to living among humans. **DISTRIBUTION** Introduced and widespread throughout Aotearoa New Zealand, but less common in the South Island. Most common around towns and cities. Rare in the southern South Island. **HABITS AND HABITATS** Can live in various habitats of damp, moist plant debris, but also even in poultry manure. Typically found in houses, grain stores and other places where grain or vegetable matter are stored and allowed to go damp. Quick life cycle and is able to produce multiple generations in a single year. Flies at night and sometimes comes to light.

PYRALIDAE & ROESLERSTAMMIIDAE

Stericta carbonalis ■ WS 14–17mm

DESCRIPTION Dark coal-coloured moth native to Australia. Dark brown or grey patterns on forewings. Hindwings lighter, with a white colour that darkens towards edges. Part of the Pyralidae family. **DISTRIBUTION** Adventive moth from Australia recently established in Aotearoa New Zealand; first recorded in 2009 on Banks Peninsula. Has already spread throughout most of the country, with records from Taupo to Otago. **HABITS AND HABITATS** Larvae feed on dead eucalypt leaves. Adults visits flowers and come to light.

Titoki Moth ■ *Vanicela disjunctella* Body Length 8–10mm

DESCRIPTION Small but striking moth with distinctive black and white colouring. Head, front of thorax and legs white. Long, hairy antennae that are often held against body, and white labial palps curve in front of head. **DISTRIBUTION** Throughout the North Island and upper half of the South Island. Closely follows distribution of its host plant. **HABITS AND HABITATS** Widespread in native forests, shrubland, parks and gardens. Probably two or more generations per year. Moths fly in October–April. Larvae leave distinctive leaf mines on Titoki leaves. Their distinctive basket-weaved cocoons may be found in fold of leaves of its host plant, the Titoki *Alectryon excelsus*, and plants growing under Titoki trees.

SATURNIIDAE & SPHINGIDAE

Emperor Gum Moth ▪ *Opodiphthera eucalypti* WS 120–150mm

DESCRIPTION Remarkably large, magnificent nocturnal moth adapted to eucalyptus trees. Native to Australia (present in all states) and an introduced species in Aotearoa New Zealand. Wings and body pale brown or reddish-brown in base colour. Wings adorned with four conspicuous, colourful eye-spots and various curved blackish and white lines. **DISTRIBUTION** Widespread in the North Island and upper half of the South Island. **HABITS AND HABITATS** Adult moths live for a few days or weeks and do not feed. Eggs usually laid on eucalyptus leaves. Larvae hatch after 7–10 days. Adult larvae large and bluish-green with a yellow lateral line, and also with blue-tipped or red-tipped protuberances along sides and back, known as scoli. When they are fully mature, they spin a protective dark brown silk cocoon on a branch. The moth usually emerges from the cocoon the following year when weather conditions are suitable (spring or early summer).

Convolvulus Hawkmoth ▪ *Agrius convolvuli* WS 80–105mm

DESCRIPTION Large hawkmoth common throughout much of the world. Mostly grey but with black and pinkish-red bands edged with white on abdomen, giving an intriguing appearance. **DISTRIBUTION** Most observations are from northern half of the North Island, and rarely reported from the South Island. Also present in Europe, Africa, Asia and Australia. **HABITS AND HABITATS** Often flies around sunset and can be seen in gardens hovering over flowers. Larvae eat the leaves of convolvulus (bindweeds), as well as a wide range of other food plants. Also feeds on sweet potato or kūmara in Aotearoa New Zealand and the Pacific. **REMARKS** Very attracted to light. Lengthy proboscis that comes in handy for feeding on long, horn-like flowers.

STATHMOPODIDAE & THYRIDIDAE

Stathmopoda spp. ■ WS 9–15mm

DESCRIPTION Small, unusual moths with narrow wings and distinctive hindlegs. When at rest, hindlegs are held out to the sides of the wings and above the body, a resting posture characteristic of this family. Hindlegs have dense whorls of hair, which gave rise to the common name of 'featherfoot' applied to some species. About 20 species in this genus in Aotearoa New Zealand, most of which are endemic, but there are also a few introduced species, such as the Eriococcus Caterpillar *S. melanochra*. Two of the more commonly encountered endemic species are pictured, the Grey-lined Featherfoot *S. plumbiflua* and Kowhai Seed Moth *S. aposema*. **DISTRIBUTION** Kowhai Seed Moth throughout the North and South Islands; Grey-lined Featherfoot in the lower North Island and throughout the South Island. **HABITS AND HABITATS** Occur in native forests, shrubland, coastal dunes, gardens and orchards. Grey-lined Featherfoot adults found throughout the year; Kowhai Seed Moth rarely seen in winter and most common in spring and early summer. Both species attracted to light or can be disturbed from vegetation by day. Larvae bore into seeds, fruits or buds of plants.

Grey-lined Featherfoot Stathmopoda plumbiflua   *Kowhai Seed Moth* Stathmopoda aposema

Muehlenbeckia Stem Gall Moth ■ *Morova subfasciata* WS 20–40mm

DESCRIPTION Moth with brownish-orange wings heavily folded near costa, providing a convincing imitation of a crumpled dead leaf. Females much larger and paler than males. Aotearoa New Zealand's only member of the Thyrididae family, a globally widespread and largely tropical moth family. **DISTRIBUTION** Throughout New Zealand, but rare in the upper North Island. **HABITS AND HABITATS** Found in native forests, shrubland, coastal dunes and gardens. The large swellings or galls on the Pōhuehue *Muehlenbeckia australis* and, more rarely, on *Parsonsia* vines, are caused by the larvae tunnelling into stems as they feed. They pupate in the stems within the gall, having first chewed a tunnel to an exit in the bark for later emergence as a moth. Adults found mostly in November–February and attracted to light.

TINEIDAE

Erechthias capnitis & *E. terminella* ■ Body Length 9mm (e)

DESCRIPTION Two small moths. *E. capnitis* originally endemic to Norfolk Island but recorded as having arrived in Aotearoa New Zealand by 1977; *E. terminella* endemic to New Zealand. **DISTRIBUTION** *E. capnitis* mostly present in upper half of the North Island, where it is widespread, but has recently been reported in New Plymouth and Wellington. *E. terminella* occurs throughout the North Island and also present in the upper South Island. **HABITS AND HABITATS** Both species on the wing throughout the year. *E. capnitis* larvae feed on dead dry wood or dry plant stems, usually above ground, and have been recorded feeding on the New Zealand Cabbage Tree *Cordyline australis*. *E. terminella* larvae feed on ripening fruits, such as Pukanui *Meryta sinclairii*, and have also been reared from dead stems of *Lupinus arboreus* and *Pittosporum* fruits. Also reared from dead bark of pear *Pyrus* spp. and dead stems of New Zealand spinach *Tetragonia* spp. Attracted to light.

Erechthias capnitis

Erechthias terminella

Dusky Scuttler ■ *Opogona omoscopa*, & *O. comptella* WS 15–17mm

DESCRIPTION Two small moths that may turn up in homes. The Dusky Scuttler has a cosmopolitan distribution, being found in many countries. It has been moved around by human trade and its original origin is uncertain. *O. comptella* found in Australia and introduced to Aotearoa New Zealand. The Dusky Scuttler is speckled brown in appearance, whereas *O. comptella* has distinctive light yellow patches over otherwise black or purplish-back forewings. **DISTRIBUTION** Both species found throughout most of the country. **HABITS AND HABITATS** Can turn up in almost any habitat type, although *O. comptella* appears restricted to somewhat drier habitats and is most common in the east of both islands, and not present in dense native forest. Both frequent gardens, dwellings and remote natural habitats, and often turn up inside houses. They fly at night and come to light. Larvae feed on a wide range of decaying vegetable matter.

Dusky Scuttler Opogona omoscopa

Opogona comptella

■ TINEIDAE ■

Dead Sheep Moth & Bird Nest Moth ■ *Monopis ethelella* & *M. crocicapitella* WS 10–16mm

Dead Sheep Moth Monopis ethelella

DESCRIPTION Two small moths with unusual common names related to the feeding habits of the larvae. The Dead Sheep Moth is native to Australia and native/self-introduced to Aotearoa New Zealand. The Bird Nest Moth has a cosmopolitan distribution and is introduced. To distinguish the two species from each other, the Bird Nest Moth generally has more light-coloured speckling on forewings, and cream stripe along dorsum is more orange, and has smoother margins, than in the Dead Sheep Moth. Also, the latter has fine orange stripe along forewing costa. The Bird Nest Moth does not have orange on costa, at most some pale yellow. **DISTRIBUTION** Both species widespread throughout New Zealand, but the Dead Sheep Moth is more commonly seen. **HABITS AND HABITATS** Both species can turn up in almost any habitat type, including around gardens and dwellings, as well as remote natural habitats. They fly at night and come to light. Larvae of *Monopis* species feed on dried animal and/or vegetable matter. Both species have been found on dead animals, including rats and sheep, possum skins, pigeon guano, chicken manure, bird's nests, stored wool, and clothing and textiles. Bird Nest Moth larvae also known to feed on stored products of vegetable origin such as flour, corn and felt. Larvae construct silk tunnels through their feeding area and pupate there. **REMARKS** Larvae of *Monopis* species are able to digest keratin in wool and feathers. There are also at least three endemic species of *Monopis* in New Zealand, but these are rarely seen. Pale, thinly scaled spot on forewings of *Monopis* has recently been demonstrated to be an 'aeroelastic tymbal', which distorts in flight to make a high-pitched noise, confusing echolocating bats. The species in the Chatham Islands, *M. typhlopa*, lacks this spot, as the Chatham Islands lack bats.

Bird Nest Moth Monopis crocicapitella

TORTRICIDAE

Apoctena flavescens & *A. conditana* ◾ WS 16–20mm 🟢

DESCRIPTION Two endemic species in the family Tortricidae, commonly known as tortrix or leafroller moths. Both species highly variable in pattern and colour, and very similar to a number of other moth species. *A. flavescens* has a little sinuation in forewing termen marked out by dark scaling; *A. conditana* is a pallid species, often speckled with well-spaced dark scales, and the common form illustrated has anvil-like dark mark on costa. **DISTRIBUTION** Both species found widely across the North and South Islands. **HABITS AND HABITATS** Larvae polyphagous (feed on a wide variety of plants). All *Apoctena* species have arboreal, or at least above the ground-feeding larvae. They construct leaf rolls. Adults of both species attracted to light. **REMARKS** The two species used to be incorporated in the genus *Planotortrix*; however, pheromonal, behavioural and adult morphological studies forced a reassessment of these genera, leading to the establishment of two new genera: *Apoctena* and *Leucotenes* (Dugdale, 1990).

Apoctena flavescens   Apoctena conditana

Bactra noteraula ◾ WS 22mm 🟢

DESCRIPTION Small, light brown endemic moth. Forewings slightly darker than hindwings. **DISTRIBUTION** Widespread throughout the North and South Islands. Also found on the Chatham Islands and Rakiura/Stewart Island. **HABITS AND HABITATS** Adults on the wing in December–April and attracted to light. Larvae tunnel into tiller bases, shoots and inflorescences of the Coastal Cutty Grass *Cyperus ustulatus*, Pingao *Ficinia spiralis* and Tall Flatsedge *Cyperus eragrostis*.

▪ TORTRICIDAE ▪

Capua intractana & '*Capua*' *semiferana* (e) ▪ WS 18–20mm

DESCRIPTION Two common tortricid moths. *C. intractana* found in Australia and introduced to Aotearoa New Zealand. '*C*' *semiferana* endemic to New Zealand and very variable in appearance, with multiple colour and pattern forms. Its genus placement requires revision. **DISTRIBUTION** Both species widespread throughout New Zealand. **HABITS AND HABITATS** *C. intractana* on the wing throughout the year; '*C*'. *semiferana* in September–May. Both species attracted to light. Larvae of *C. intractana* have been found in leaf litter eating dead eucalyptus leaves, and based on distribution must feed on leaf litter of other plants as well. '*C*'. *semiferana* larvae feed on leaves of rosette-forming herbs (often dead or dying leaves), and the Smooth Hawksbeard *Crepis capillaris* is thought to be a host.

Capua intractana

'Capua' semiferana

Catamacta gavisana ▪ WS 14–18mm (e)

DESCRIPTION Endemic moth that is highly variable in appearance. Can have startling colours and patterns, but other individuals appear very plain. Often has bright white patch on forewing costa, but this is less visible, or absent, in some individuals. **DISTRIBUTION** Widespread throughout the North and South Islands. **HABITS AND HABITATS** Recorded throughout the year, but more commonly in warmer months. Attracted to light. Larvae feed on leaves of a wide range of shrubs and trees in genera such as *Olearia*, *Coprosma*, *Leptospermum* and *Melicytus*, as well as many others. **REMARKS** The *Catamacta* genus is endemic to Aotearoa New Zealand and has five described species.

▪ TORTRICIDAE ▪

Lucerne Leafroller Moth & Yellow Field Bell Moth
Clepsis divulsana & *C. leucaniana* (e) WS 15mm

DESCRIPTION Both species are small, yellowish moths, rather bell-shaped in appearance, and rest with wings almost flat. The Lucerne Leafroller Moth is self-introduced from Australia; the Yellow Field Bell Moth is endemic to Aotearoa New Zealand. Forewing ground colour may not differ substantially between the two species. Presence of a dark fascia or at least a dark smudge near tornus of forewing more or less diagnostic of the Lucerne Leafroller Moth. In addition, it has grey hindwings v whitish fawn hindwings in the Yellow Field Bell Moth. **DISTRIBUTION** Both species widespread throughout the North and South Islands. The Yellow Field Bell Moth is also found on the Chatham Islands. **HABITS AND HABITATS** Both found in modified plant communities (pasture or cropping areas, coastal sites, urban gardens and parks), and can be polyphagous on a wide range of indigenous and introduced herbaceous plants. Both species can be disturbed from vegetation, taking flight by day, and will also come to light at night.

Lucerne Leafroller Moth Clepsis divulsana

Yellow Field Bell Moth Clepsis leucaniana

Lyre Moth ▪ *'Cnephasia' jactatana* WS 16–20mm (e)

DESCRIPTION Named for its markings, which resemble a lyre – a stringed instrument used especially in ancient Greece. Endemic and common throughout Aotearoa New Zealand. Unlikely to be confused with any other species. **DISTRIBUTION** Widespread throughout the North and South Islands. However, mostly only in coastal sites, or near coastal sites, in the southern half of the South Island. Also present on Rakiura/Stewart Island. **HABITS AND HABITATS** Eggs laid on top side of a leaf. Larvae often found on

the Hound's Tongue Fern *Microsorum pustulatum* in a silken tube, feeding on the leaves. They also feed on a large range of other vegetation such as Kahikatea *Dacrycarpus dacrydioide*, New Zealand Flax *Phormium tenax*, kiwifruit, citrus, hawthorns, persimmons, *Eucalyptus* spp., *Fuchsia* spp. and grapevines. They primarily feed on the leaves. Adults attracted to light.

TORTRICIDAE

Cotton Tipworm Moth ■ *Crocidosema plebejana* WS 12–16mm

DESCRIPTION Widespread moth found today over much of the world but has probably been accidentally introduced to much of its current range by humans. Males dark brown to black with white dorsal patch. Females pale brown to tan with dark basal patch that does not extend to costa. There is a very similar unnamed endemic species with larvae on amaranths (Chenopodiaceae); it is usually smaller and less contrastingly marked. **DISTRIBUTION** Widespread in the North and South Islands, except for Southland and Fiordland. **HABITS AND HABITATS** Larvae polyphagous and feed inside seed capsules and shoots of a range of plants, mostly in the mallow family (Malvaceae). Adults active in the evening and attracted to lights.

Cryptaspasma querula ■ WS 21–28mm e

DESCRIPTION Endemic moth found throughout the country. Adults variable in both the appearance of yellow markings on their wings and in their size. Female tends to be more plainly coloured than male. **DISTRIBUTION** Widespread throughout the North and South Islands, but appears to be more common in the North Island. **HABITS AND HABITATS** Inhabits podocarp and broadleaved forests. Larvae consume fruits and seeds of Tawa *Beilschmiedia tawa*, Taraire *B. taraire* and Miro *Prumnopitys ferruginea*. On the wing throughout the year, and nocturnal but attracted to light.

On the left is a mating pair with the male to the left.

◾ TORTRICIDAE ◾

Leafrollers ◾ *Ctenopseustis obliquana* & *C. fraterna* WS 20–25mm

DESCRIPTION Colouration and markings on forewings very variable in both species. They range from fawn, dull-orange, chocolate-brown purplish, to brownish-white colour pattern, with or without strong overlying dark markings. Common name Black-headed Leafroller can be applied to either species. The two species are very difficult to distinguish from each other. They can only be told apart on male hindwing characteristics or on female pheromones. **DISTRIBUTION** *C. obliquana* widespread throughout the North, South and Rakiura/Stewart Islands. *C. fraterna* primarily known from the North Island, where it is widespread, but there have been some recent reports in the upper South Island. **HABITS AND HABITATS** Larvae of *C. obliquana* feed on a range of plants, including various broadleaved and coniferous trees and ferns. They feed on the leaves, stems and buds beneath a protective webbing of silk and foliage. Larvae of *C. fraterna* feed on ferns such as the Silver Fern *Cyathea dealbata*, Kātote *C. smithi*, Whekī-Ponga *Dicksonia fibrosa*, Wheki *D. squarrosa* and Umbrella Fern *Sticherus cunninghamii*. Larvae create a silken shelter by tying together the fronds of the host. In both species adult moths on the wing all year round and attracted to light.

Ctenopseustis obliquana   Ctenopseustis fraterna

Gorse Pod Moth ◾ *Cydia succedana* WS 11–16mm

DESCRIPTION European moth, introduced to Aotearoa New Zealand for biological control of the highly invasive weed Gorse *Ulex europaeus*. Forewings mostly glossy greyish with brown markings and series of silvery-white striations along costal edge. **DISTRIBUTION** Widespread throughout the North, South, Rakiura/Stewart and Chatham Islands. **HABITS AND HABITATS** Larvae feed on seeds of Gorse *Ulex europaeus*, Brooms *Cytisus scoparius*, and *Lotus* spp. Attracted to light.

◾ TORTRICIDAE ◾

Dipterina imbriferana ◾ WS 15mm (e)

DESCRIPTION Attractive silvery-grey or bronze-white endemic moth. Curved white band across centre of forewings. Dark patches or blotches at central inner edge of forewings. **DISTRIBUTION** Widespread throughout the North Island and eastern side of the South Island. **HABITS AND HABITATS** On the wing mostly in October–April. Larvae tunnel into fruits of Māhoe *Melicytus ramiflorus*. Fully grown larva dull olive-green. When ready it leaves the fruit and bores into dry wood or under bark, where it forms a cocoon and pupates.

Sharp-tipped Bell Moth ◾ *Epalxiphora axenana* WS 20–30mm (e)

DESCRIPTION Unusual-looking endemic moth with strong sexual dimorphism. Wing pattern highly variable. To assist with identification in the field, check the variation within the species on the website iNaturalist (p. 173). Also note that this moth has an unusual wing shape and resting pose that can help distinguish it from other species. Adult has strange tufts on thorax that lend support to a twig-like camouflage. **DISTRIBUTION** Throughout the North Island and the Nelson, Buller and Westland districts of the South Island. **HABITS AND HABITATS** Occupies native forest. Larvae feed on a wide range of forest trees and shrubs, such as the Broadleaf *Griselinia littoralis*, Toro *Myrsine salicina*, Kawakawa *Piper excelsum* and Kohekohe *Didymocheton spectabilis*. They seem to prefer larger leaved trees and shrubs and retreat between spun leaves. A larva pupates in or near the feeding place. Adults on the wing throughout the year but most commonly seen in summer. They rest on vegetation by day and fly at night. Attracted to light.

Male

Female

TORTRICIDAE

Light Brown Apple Moth ■ *Epiphyas postvittana* WS 16–25mm

DESCRIPTION Australian species first recorded in Aotearoa New Zealand in 1891 and now abundant nationwide. Adults light brown, yellowish moths with varying amounts of darker brown. Females larger than males, and usually have less distinct markings. **DISTRIBUTION** Widespread throughout the North and South Islands. **HABITS AND HABITATS** Occurs in a wide range of habitats, including parks, gardens, weedy areas, coastal dunes, orchards, plantations, shrubland and forest edges. Adults found throughout the year but most commonly in October–April. They are attracted to light and can also be disturbed from vegetation by day. Larvae feed on a massive variety of plants, including numerous crops. In New Zealand, more than 250 host species recorded. Feeds on nearly all types of fruits, ornamentals, vegetables, glasshouse crops and even young pine seedlings.

Eurythecta zelaea ■ WS 10mm

DESCRIPTION Delightful little endemic leaf roller of dry herb fields and salt pans in the inland South Island. In strongly marked individuals, there are distinctive dark brown or black chevron stripes over a rich orange-red base colour, which is speckled with white. Other individuals less markedly coloured and patterned, and can be whitish, usually still with one or more darker chevrons. **DISTRIBUTION** Endemic to dry parts of Otago and Canterbury. Central Otago and the Mackenzie Basin are strongholds. **HABITS AND HABITATS** Adult males fly by day low to the ground over herb fields, grassland and saltpans. Adult females brachypterous (flightless). Larvae feed on a wide range of plants. Adult males on the wing in September–April.

TORTRICIDAE

Harmologa amplexana ■ WS 16–18mm ⓔ

DESCRIPTION Endemic whitish-grey and black moth with classic 'bell shape' of many Tortricidae, giving the family its alternative English name, 'bell moths'. **DISTRIBUTION** Widespread from the Central North Island south to Otago. In the South Island, only found east of the main divide. **HABITS AND HABITATS** On the wing in September–April. Larvae web together and feed on dead leaves. Known hosts are the Pōhuehue *Muehlenbeckia complexa* and *Olearia odorata*. Attracted to light and can also be disturbed from vegetation by day.

Orange Fruit Borer ■ *Isotenes miserana* WS 20mm

DESCRIPTION Australian species introduced to Aotearoa New Zealand. Adult moths patchy speckled grey, sometimes with small patches of orange scales. Often with a broad, indistinct dark diagonal stripe across each forewing. **DISTRIBUTION** Widespread throughout the North Island, but mostly found in the upper half of the North Island.

HABITS AND HABITATS Eggs laid in a mass protected by a barrier of scales deposited around them by the female moth. Larvae polyphagous, feeding on a wide range of flowers and fruits of a large variety of agricultural plants and fruit trees, including avocadoes, grapes, oranges and macadamia. Larvae were first found in New Zealand on dead petals and leaves of *Camellia* spp.

■ TORTRICIDAE ■

Leafrollers ■ *Planotortrix* spp. WS 15–32mm e

DESCRIPTION Endemic leafrollers strongly variable in colour and pattern, sometimes having diamond-shaped patch on forewings. The Blacklegged Leafroller *P. notophaea* can be identified based on smaller size and black spots or black line along sides of abdomen; *P. excessana* cannot be distinguished from *P. octo* except on the basis of sex pheromone of females. **DISTRIBUTION** Widespread on the South, North and Chatham Islands. **HABITS AND HABITATS** *Planotortrix* moths live in a wide variety of habitats. Larvae polyphagous and feed on many forest, orchard and garden shrubs and trees. Endemic host species for the Blacklegged Leafroller include narrow-leaved hosts like *Podocarpus*, *Kunzea*, *Leptospermum* and *Ulex*, as well as the Matai *Prumnopitys taxifolia*. *P. excessana* utilizes endemic hosts such as *Coprosma lucida*, *Griselinia* spp. and new foliage of *Melicytus ramiflorus*, as well as exotic hosts including species of *Camellia*, *Hedera*, *Prunus* and *Pyrus*. Larvae, like many leafrollers, construct a shelter made of foliage webbed together with silk. Can be on the wing throughout the year, and attracted to light.

Blacklegged Leafroller Planotortrix notophaea   *Either* Planotortrix excessana *or* P. octo

Painted Wedge ■ *Pyrgotis plagiatana* WS 18mm e

DESCRIPTION Variable endemic species, differing greatly in intensity of its colouring and markings. Often a highly contrasting pattern, including white and rich reddish-brown. **DISTRIBUTION** Widespread throughout the North, South and Rakiura/Stewart Islands. **HABITS AND HABITATS** Larvae polyphagous. They web together and feed on leaves and stems of various trees and shrubs, including (but not limited to) New Zealand brooms *Carmichaelia* spp., Kōtukutuku *Fuchsia excorticata*, *Veronica* shrubs, Poataniwha *Melicope simplex*, *Metrosideros* spp., Tauhinu *Ozothamnus leptophyllus*, Kōhūhū *Pittosporum tenuifolium*, *Pinus radiata* and Gorse *Ulex europaeus*. Flies at night and attracted to light. On the wing year round where climate is warm enough.

XYLORYCTIDAE

Gymnobathra spp. ■ WS 13–20mm (e)

Gymnobathra sarcoxantha

DESCRIPTION Genus of 23 described species endemic to Aotearoa New Zealand; may not form a natural group and several genera may need to be recognized in future. Five of the most commonly encountered species are profiled: the Pink-tipped Yellow Moth *G. flavidella*, *G. sarcoxantha*, *G. parca*, *G. tholodella* and the Small Angle-wing Moth *G. hyetodes*. In the Pink-tipped Yellow Moth forewings are bright yellow, with pinkish and black; *G. sarcoxantha* pale straw to reddish-brown, usually with two dots and comma-like mark on each forewing, but sometimes these are faint or absent; *G. parca* similar to *G. sarcoxantha* but generally more orange (or darker) in colour and has shorter and broader wings; *G. tholodella* light cream in colour often heavily sprinkled with dark brown. The Small Angle-wing Moth is easily recognizable based on its 'sickle-shaped' forewings. Male rather dull brown, but female bright reddish-brown in this sexually dimorphic species.

DISTRIBUTION The Pink-tipped Yellow Moth widespread in the North Island and in the South Island from Canterbury northwards. It is likely that true *G. sarcoxantha* is confined to the southern South Island. All records from further north likely to belong to an unnamed species with dark-clouded hindwings. *G. parca* present in the lower North Island and eastern side of the South Island. *G. tholodella* widespread on the North and South Islands. Small Angle-wing Moth only found in the North Island, where it is widespread.

Pink-tipped Yellow Moth Gymnobathra flavidella

XYLORYCTIDAE

Gymnobathra tholodella

HABITS AND HABITATS

On the wing primarily over warm months of the year, with some species being more abundant in spring and early summer and others later in summer and autumn. All species fly at night and come to light. Some also fly by day or can easily be disturbed from vegetation. G. *sarcoxantha*, G. *parca* and G. *tholodella* have case-making larvae feeding on leaf litter. G. *hyetodes* and G. *flavidella* have larvae in dead wood; the Small Angle-wing Moth has been reared from dead twigs of *Melicope ternata*, and the Pink-tipped Yellow Moth from dead twigs of Rangiora *Brachyglottis repanda* and dead stems of *Gahnia procera*.

Gymnobathra parca

Small Angle-wing Moth Gymnobathra hyetodes

XYLORYCTIDAE

Lichen tuft moths ■ *Izatha* spp. WS 13–29mm ⓔ

DESCRIPTION This remarkable genus has been popularly named the 'lichen tuft moths'. Many *Izatha* species are stunningly camouflaged when resting on trunks of forest trees and shrubs. This is in part due to the white-and-black disruptive patterning that many possess, but also to the tufts of raised scales on their wings, which mimic the structures of lichens. The genus is a diverse group of endemic moths, with 40 species recognized. *I. acmonias*, *I. convulsella*, *I. huttonii* and *I. katadiktya* are profiled here as examples. **DISTRIBUTION** *I. acmonias* and *I. katadiktya* widespread in the South Island. *I. convulsella* widespread from the central North Island to Southland. *I. huttonii* widespread in the South Island and also occurs in Wellington. **HABITS AND HABITATS** Larvae of most *Izatha* species tunnel into dead rotting wood, where they probably digest the fungal element. Larvae of other species have been found in bracket fungi or are known or suspected to feed on lichens

Izatha katadiktya

Izatha huttonii

(and/or epiphytic mosses). Thus, *Izatha* species form part of the country's decomposer community, which is fundamental for recycling nutrients in forests and shrubland. *Izatha* species are on the wing primarily over warm months of the year and the most common species (and some rarer ones) are attracted to light. Some *Izatha* can be found by day by searching among lichens for resting moths. **REMARKS** Male *Izatha* are noted for their strange and extreme genitalia. The phallus (penis) is often endowed with strong ridges bearing backwards pointing teeth, and damage presumed to be from these teeth has been seen in the female's genital tract. Some females have the genital tract reinforced, presumably to minimize damage during mating. The 'phallus-teeth' may have evolved as a way for the male to discourage the female from mating more than once.

Izatha convulsella

Izatha acmonias

ZYGAENIDAE

Bamboo Moth ■ *Artona martini* WS 24mm

Larva

DESCRIPTION Asian moth, native to China and Japan. First found in Whangarei in 1996. Its introduction represented the first record of Zygaenidae in Aotearoa New Zealand, as there are no native moths in this family. Both the moths and larvae are highly distinctive. Moths metallic bluish-black. Larvae feed on various bamboo species and are pale yellow with large black spots and large urticating hairs. These can cause a skin rash accompanied by a burning sensation, so handling the caterpillars should be avoided. Caterpillars gregarious in their younger stages. Pupation occurs in a pale brown, papery cocoon. **DISTRIBUTION** Only known from the Northland, Auckland, Waikato and Bay of Plenty regions. **HABITS AND HABITATS** Larvae can occur in very high numbers, although they can be quite rare between outbreaks. In the late 1990s cast caterpillar skins and dead pupae of the moth were intercepted on secondhand vehicles imported from Japan at Auckland, Tauranga and Nelson. This suggests a likely mode of entry into New Zealand. **REMARKS** There are more than 1,000 species in the Zygaenidae worldwide. They are often colourful moths and many use toxic chemicals from their host plants to make adults distasteful or poisonous to predators.

Adult moth

RARITIES & SPECIALITIES

To balance the emphasis on common species, below are profiles of a selection of rare or particularly ecologically fascinating Lepidoptera found in Aotearoa New Zealand.

Western Tiger Moth ■ *Metacrias erichrysa* WS 31–33mm

DESCRIPTION Named for the male's spectacular bright orange and black markings, reminiscent of a tiger. Endemic and in the Erebidae family. Adults males brightly coloured, diurnal and strong fliers. In contrast, females are flightless, buff coloured and very short winged. Larvae typically 2–3cm long, black and hairy, with some bright blue or orange colouring. Forewing orange colour same as hindwing, whereas in M. *huttoni* forewings are paler than hindwings. There are three endemic *Metacrias* species in the South Island with a vaguely similar ecology. **DISTRIBUTION** Known from the Ruahine and Tararua Ranges in the North Island. Widespread on the western side of the South Island, including the Queenstown Lakes area (particularly subalpine or alpine areas). **HABITS AND HABITATS** Inhabits open herb and tussock fields in mountainous terrain, primarily at altitudes of 900–1,700m. Males fly rapidly by day and attracted to females by pheromones. Females flightless and tend to stay in their cocoons to breed and lay eggs. Once hatched, larvae feed on corpse of female before dispersing widely and feeding on a wide range of tussocks, grasses, flowers, vines and herbs. Since females cannot fly, and main population dispersal is as a result of larval movement. Male on the wing in mid-November–March. **REMARKS** For those with a keen eye, *Metacrias* moths can be seen during mild weather in the alpine zone flying over tussock grassland or rocky terrain. They appear as very fast-moving, orange-and-black 'blurs'.

▪ Rarities & Specialities ▪

Otago Ghost Moth ▪ *Aoraia rufivena* WS 60–74mm 🟢

DESCRIPTION Impressively large endemic species in the Hepialidae family. Male wingspan 60–74mm. Forewing ground colour pale and dark brown, with ash-white pattern. Females subbrachypterous (partly flightless) with a wingspan of 55–68mm. Adult males on the wing from January to early winter, seeking out short-winged, flightless females. **DISTRIBUTION** Predominantly southeastern distribution. Mainly found in Otago but edging into South Canterbury and Southland. **HABITS AND HABITATS** Occurs in native forests and high-country shrubland, tussockland and swampy habitats. Adult males readily come to light. **REMARKS** Larvae known hosts of the fungus known as the Vegetable Caterpillar *Ophiocordyceps robertsii*. This is an entomopathogenic fungus (one usually attached to the bodies of insects). It enters larvae that live in the leaf litter and preserves them to act as a base for the fungus to grow. Subsequently, the fungus sends a fruiting spine from the caterpillar's preserved body up above the leaf litter and releases its spores. Caterpillars accidentally eat the spores when feeding on leaf litter and the fungal life cycle continues.

▪ RARITIES & SPECIALITIES ▪

Sphagnum Porina ▪ *Heloxycanus patricki* WS 40–55mm (e)

DESCRIPTION Special moth with a fascinating ecology and the only member of the endemic genus *Heloxycanus*. Females larger than males (48–55mm v 40–45mm). Southern populations generally brown with diminished forewing stripe. Populations in northern part of range yellowish-fawn with characteristic white stripe on forewings (as pictured from Danseys Pass). **DISTRIBUTION** Otago Lakes area, Central Otago including Danseys Pass, Dunedin, Southland, Fiordland and Stewart Island. **HABITS AND HABITATS** Found in damp, swampy habitats such as mossy bogs and in *Sphagnum* bogs in southern parts of the country. Late-autumn moth that begins to emerge in late March or April, and usually finishes around early–mid June. Unusual two-year life cycle. High numbers of adults only occur in odd-numbered years. Larvae might feed on moss rhizoids. Pupal cases can sometimes be located sticking out of the moss and are a useful indicator that the species is present. **REMARKS** Named after its discoverer, Brian Patrick, who first found the species at Danseys Pass in April 1979. Vulnerable to loss of alpine wetland habitat and sphagnum bogs in southern Aotearoa New Zealand.

▪ Rarities & Specialities ▪

Aponotoreas dissimilis ▪ WS 34–40mm 🟢

DESCRIPTION Beautiful endemic moth in the Geometridae family. Colour varies from greyish-brown to reddish-brown. Forewings a unique elongated triangle shape in both sexes, differing substantially from other *Aponotoreas* species. Distinctive feature is very pointed apex of hindwings. **DISTRIBUTION** Widespread in alpine and subalpine areas across the whole of the South Island, but rarely seen and probably sparse. Most records close to the Southern Alps or in western mountainous areas such as Fiordland, South Westland and western Southland. **HABITS AND HABITATS** Host plants are *Dracophyllum* shrubs, which grow in mountainous habitat. **REMARKS** Unusual in being the only moth in the *Aponotoreas* genus that is nocturnal and readily comes to light. Also unusual for its wide wingspan and peculiar wing shape.

Wirerush Looper ▪ *Aponotoreas synclinalis* WS 24–28mm 🟢

DESCRIPTION Stunningly marked endemic day-flying moth in the Geometridae family. Forewings characterized by strongly contrasting, streaked markings in black, white, purple-grey, and sometimes bright orange or yellow. **DISTRIBUTION** Rarely encountered moth of the deep south but can be locally abundant. Only known from Southland, Stewart Island and South Otago (high swampy areas of the Catlins). Occurs in wetlands around the southern coast of Southland, including Tiwai Point, Seaward Moss Conservation Area and Slope Point. **HABITS AND HABITATS** Day flying and on the wing in October–March. Larvae feed on Wirerush *Empodisma minus*. In addition, uses *Dracophyllum politum* in high-elevation parts of Stewart Island.

▪ Rarities & Specialities ▪

Exquisite Carpet ▪ *Asaphodes adonis* WS 25–28mm

DESCRIPTION Stunning green (or blue-green) and black moth in the Geometridae family. Blueish-green or whitish-green forewings, typically with black, narrow to broad wavy lines edged with white. Hindwings usually have characteristic pale rose tinge. **DISTRIBUTION** Endemic to the South Island. Regarded as uncommon. Known sites include Lewis Pass, Castle Hill, Westland, Fiordland, Otago Lakes area, Takitimu Mountains and Wakaia Bush, Southland. **HABITS AND HABITATS** Found in native forest at 300–1,200m. Adults on the wing in November–March (peak in mid-summer). Will come to light. Larvae have been reared on *Ranunculus* spp. (buttercups).

Asaphodes ida ▪ WS 20–26mm

DESCRIPTION Rare endemic looper moth. Head, palpi, abdomen and thorax grey. Forewings elongated triangular, and brownish-grey with mixture of reddish and white crenulate lines. Blackish discal spot. Hindwings light–moderate grey-brown. Distinguished from outwardly similar *Hydriomena hemizona* based on presence of antennal pectinations. Also somewhat similar in appearance to *Austrocidaria cedrinodes* (p. 46), but pattern differs. **DISTRIBUTION** Discovered at Eweburn Stream, Ida Range. Only recorded from half a dozen localities in Central Otago (Rough Ridge, Saint Bathans area, Ida Range, Hawkdun Range, Danseys Pass and the Manuherikia River catchment). May occur in the Mackenzie Country. **HABITS AND HABITATS** Has been found in subalpine or alpine wetland habitat at 800–1,100m. Associated with *Ranunculus* (buttercups). Adults on the wing in February and March. Will come to light.

▪ Rarities & Specialities ▪

Yellow and Brown Carpet ▪ *Asaphodes prasinias* WS 22–26mm (e)

DESCRIPTION Vividly coloured endemic moth in the Geometridae family. Bright yellow and orange with numerous cloudy reddish-brown or purplish-brown bands across forewings. **DISTRIBUTION** Widespread throughout the South Island and central-lower North Island, but very sparse and rarely encountered. Most often recorded at inland or subalpine sites, but rare coastal populations occur in Southland and Otago. **HABITS AND HABITATS** Inhabits native forest including beech forest and subalpine scrub; also often among tussock grassland. Recorded up to at least 1,600m. Adults on the wing in November–January. Larvae have been raised on buttercup species in the *Ranunculus* genus. Day flying, but also frequently comes to light at night. **REMARKS** Has disappeared from some former sites.

Rusty Hebe Looper ▪ *Dasyuris callicrena* WS 32–36mm

DESCRIPTION Stunning black, white and orange-striped, day-flying moth. Forewings predominantly black, usually with patches of orange scales and a number of broad or narrow crenulate white bands. Hindwings similar in colour and pattern to forewings but paler. Undersides of wings bright orange-yellow, with sparse black patches and white bands, similar to those on dorsal surface. **DISTRIBUTION** Throughout the South Island from northern parts of Kahurangi National Park to southern Fiordland but very sparse. Most common in mountainous parts of Otago and Fiordland. Typically recorded in subalpine or alpine zone at 800–1,600m. **HABITS AND HABITATS** Adults day flying and on the wing in November–February. Host plants are in the genus *Veronica* (formerly *Hebe*). **REMARKS** Feature of *Hebe odora* shrubland among tussock grassland or wetlands in hilly or mountainous parts of the South Island. Fast flier.

■ RARITIES & SPECIALITIES ■

Declana nigrosparsa ■ WS 32–36mm e

DESCRIPTION Spectacular endemic looper moth in the Geometridae family. Good example of how the same moth species can look markedly different from one location to the next. Two forewing patterns. Form '*nigrosparsa*' evenly pale grey on forewings and weakly marked in darker grey with faint transverse lines. Form '*toreuta*' (previously regarded as a separate species) much more boldly marked in creamy-white, with sharply defined dark grey and black markings.

DISTRIBUTION Form '*toreuta*' widespread, but rare, in the central North Island and the South Island. Form '*nigrosparsa*' only known from eastern side of the Southern Alps (Banks Peninsula southwards). At Arthur's Pass in Canterbury, both forms occur together. **HABITS AND HABITATS** Comes to light, especially in winter, spring and early summer. Both forms feed exclusively on small-leaved species of host genus *Olearia*, such as *O. odorata*. '*Nigrosparsa*' form is the only form found in drier and exposed grey shrubland communities on the eastern side of the Southern Alps, where it adopts a plain grey colour, possibly to help it blend in with greyish divaricating shrubs. Form '*toreuta*' occupies high-rainfall forests and subalpine shrubland. **REMARKS** The two forms were recently confirmed by taxonomists to be the same species, despite vast differences in forewing colour and pattern. This was partly based on the male and female genitalia of both forms being identical to each other. Considered threatened due to loss of its host communities (grey shrubland or forest edges with an *Olearia* component) to increasingly intensive farming practices and the supreme ease of shrubland destruction.

▪ Rarities & Specialities ▪

Dichromodes spp. ▪ WS 16–24mm

DESCRIPTION Genus of moths common in Australia and Aotearoa New Zealand; however, New Zealand species require taxonomic work and may belong in a separate genus. Six described endemic species in New Zealand but there are known to be another half a dozen or more undescribed species. Described species are: *D. sphaeriata, D. gypsotis*, the Blue & Orange Rock Looper *D. ida, D. niger, D. cynica* and *D. simulans*. **DISTRIBUTION** *Dichromodes* are known from rocky regions of the South Island, particularly Central Otago and the Canterbury high country, as well as Banks Peninsula and the Port Hills near Christchurch. Sparse in the North Island, aside from Wellington area. **HABITS AND HABITATS** Swift, day-flying moths of rocky habitats such as tors, bluffs, outcrops, gullies, screes and boulder fields. Larvae feed on various lichen species that grow on and around their rocky habitat. In New Zealand, only *D. sphaeriata* is known to come to light. **REMARKS** Some of the undescribed entities are considered quite rare. For

Dichromodes simulans

Dichromodes gypsotis

example, *Dichromodes* 'Gore Bay' has been ranked by the Department of Conservation as Threatened-Nationally Critical. It is tricky to attempt to find and photograph these swift day-flying moths. For example, the Blue and Orange Rock Looper lives on large rock tors in Central Otago and on even the quietest approach, tends to launch itself swiftly into the air and fly tens of metres before resettling gently on another rock tor.

Blue & Orange Rock Looper Dichromodes ida

Dichromodes sphaeriata

= RARITIES & SPECIALITIES =

Streaked Inanga Looper ■ *Ipana glacialis* WS 27–35mm

DESCRIPTION Dazzling moth from the alpine zone of the South Island. Stands out among the sister genera of *Ipana* and *Declana* (informally known as the 'declanoids'), being the only truly alpine specialist of this group, and the only species whose males fly by day. All of the other 'declanoids' predominantly fly at night at lower elevations. Unlikely to be confused with any other moth species. **DISTRIBUTION** Reasonably widespread in mountainous parts of the central and western South Island, particularly Kahurangi National Park, Westland, the Canterbury High Contry and Fiordland. Despite this, rarely reported. Very fast flier during sunny weather, and thus difficult to observe or photograph. This, in addition to occupying vast, remote and often hostile mountainous terrain, may somewhat explain the scarcity of records. **HABITS AND HABITATS** On the wing in mid-November–late January (early midsummer). Host plants are small-leaved shrubs in the *Dracophyllum* genus (inaka).

■ RARITIES & SPECIALITIES ■

Notoreas elegans ■ WS 20–24mm

DESCRIPTION Elegant endemic diurnal moth in the Geometridae family. Adult moths have bright orange, black and white markings on forewings with orange-and-black-marked hindwings. **DISTRIBUTION** Low elevations to 1,900m in lowland to alpine areas of the Canterbury High Country, Mackenzie Basin, Central Otago, West Otago and Fiordland. **HABITS AND HABITATS** Larval host plants are endemic species in the genus *Pimelea* (rice flowers). Female lays her eggs within the flower buds. Adults are day-flying moths of open tussock grassland, shrubland, forest edge and wetland habitats. Fast but low flyers. **REMARKS** Larvae of *Notoreas* moths feed entirely on the daphne family Thymelaeaceae, which is represented in Aotearoa New Zealand by the genera *Pimelea* and *Kelleria*.

Snowberry Yellow ■ *Orthoclydon chlorias* WS 30–34mm

DESCRIPTION Bright yellow looper moth and one of three endemic species in the genus *Orthoclydon*. Forewings bright yellow with series of purplish spots. May or may not contain one or more larger dark purple smudges on forewings near costa. Hindwings similar, but paler than forewings. Can be confused with the Aristotelia Looper *Epiphryne xanthaspis*, but this has a distinctive purplish-brown mark halfway along costa. **DISTRIBUTION** Widespread in the South Island but rarely encountered, and predominantly found in central and western areas from Kahurangi National Park to Fiordland. **HABITS AND HABITATS** Larvae feed on leaves of *Gaultheria* spp., such as the Mountain Snowberry G. *depressa*, hence the common name Snowberry Yellow. **REMARKS** The *Orthoclydon* genus also contains the common Flax Window-maker (p. 63) and the very rare *O. pseudostinaria*, which is classified as Critically Endangered by the Department of Conservation.

▪ Rarities & Specialities ▪

Paranotoreas fulva ▪ WS 16–20mm (e)

DESCRIPTION Endemic moth in the Geometridae family. Forewings dull greyish-brown, sometimes greenish tinged. Wavy dark bands across forewings. Hindwings dull orange-brown or reddish-brown. Individual moths vary in intensity of colouring. **DISTRIBUTION** Central Otago and South Canterbury, including Tekapo, the Maniototo, Alexandra, Wanaka, Hawea and Tarras. **HABITS AND HABITATS** Known to inhabit salt pans in Otago and glacial terraces in the Mackenzie Country. Also persists at some depleted sites without saline vegetation (which were historically saltpans). At degraded sites has adapted to exotic herbs to survive. Larvae primarily feed on endemic groundcover *Atriplex buchananii* and introduced *Plantago coronopus*. Adults day flying in October–December and March. **REMARKS** Often seen sunbathing on bare, salty soil in their favoured habitats. Protection of inland saltpans is important for the conservation of this special moth.

Pseudocoremia dugdalei ▪ WS 28–34mm (e)

DESCRIPTION Rarely encountered endemic looper. Alike in appearance to a few other *Pseudocoremia* moths, such as the Brown Forest Flash (p. 71), but slightly smaller and

broad, straight white fascia that is not interrupted or narrowed above mid-point is characteristic of this species. **DISTRIBUTION** Elusive species known from very few sites, but widely spread out across the country. Known sites include the Waitakere Ranges near Auckland, Nelson region (for example Golden Bay), and Dunedin area Otago Peninsula (Orokonui Ecosanctuary and Sawyer's Bay). **HABITS AND HABITATS** Comes to light. On the wing in December–March. Life history and larva unknown.

▪ RARITIES & SPECIALITIES ▪

Spindle moths ▪ *Tatosoma* spp. WS 28–38mm

DESCRIPTION Spindle moths (*Tatosoma*) in the Geometridae family are found only in Aotearoa New Zealand. There are nine described species. See the Kāmahi Green Spindle (p. 74); this section covers the other eight described species. Males of all *Tatosoma* moths have a distinctive thin, elongated abdomen that stretches well beyond wings when moth is at rest. In females, abdomen ends approximately at termen of forewings, which are often heavily marked with greens, browns and sometimes reds. Hindwings generally lightly marked or not marked, and much paler. Some species, such as the Tutu Green Spindle *T. lestevata*, have a highly distinctive pattern, unlike others in the genus. Others are more difficult to tell apart and may require expert opinion, but with practice many of the subtle differences between species can be learnt.
DISTRIBUTION Throughout the country, but generally most common in areas with extensive

Tutu Green Spindle Tatosoma lestevata

Tutu Green Spindle Tatosoma lestevata

Tatosoma alta

Tatosoma fasciata

▪ Rarities & Specialities ▪

Tatosoma monoviridisata

Tatosoma agrionata

Tatosoma transitaria

Tatosoma topea

native forests, such as beech, podocarp or broadleaved forest. Most species somewhat sparse in occurrence and uncommonly seen, but generally have very wide distributional ranges. *T. monoviridisata* is a perfect example, being present in both main islands, but only recorded from a very small number of sites – the Tararua Ranges, Tongariro National Park, Waitati near Dunedin, Oteake Conservation Park in North Otago and Fiordland National Park. **HABITS AND HABITATS** Mostly on the wing over warmer months of the year. Host plants include Tutu *Coriaria* spp. for *T. lestevata*; Loranthaceae (mistletoes such as *Ileostylus micranthus*) for *T. agrionata*; Silver Beech *Lophozonia menziesii* for *T. fasciata*; *Phyllocladus* for *T. alta*; and Mataī *Prumnopitys taxifolia* for *T. topea*. For other species, such as *T. monoviridisata*, host plant or plants are unknown. Adults of all species are nocturnal and will come to light.

■ RARITIES & SPECIALITIES ■

Theoxena scissaria & *Theoxena* sp. 'non-pectinate' ■
WS 23–25mm

Theoxena scissaria

DESCRIPTION The endemic genus *Theoxena* in the Geometridae family has two species, *T. scissaria* and the undescribed and seemingly very rare *T.* sp. 'non-pectinate'. They are dull white or pale greyish-brown, lightly sprinkled with dark brown or black spots. In *T. scissaria* only, there is a curved black median streak, sharply defined above and suffused beneath. **DISTRIBUTION** Both species only known from Central Otago and the Mackenzie Country in areas such as Tekapo, several sites on the Maniototo, Danseys Pass, Rock and Pillar Range, Alexandra, Wanaka, Hawea and Tarras. *T.* sp. 'non-pectinate' known from far fewer sites. **HABITS AND HABITATS** Life history of *Theoxena* moths largely unknown, but possible associations with the tussock *Poa cita* and Aotearoa New Zealand brooms (genus *Carmichaelia*) have been suggested. Female *Theoxena* also unknown and may be flightless, which would help explain the lack of records of them. Both taxa come to light. *T. scissaria* known to emerge from late autumn right through winter, and has also been recently recorded every month between September and January. **REMARKS** *T. scissaria* classified as Threatened–Nationally Vulnerable by the Department of Conservation, and *T.* sp. 'non-pectinate' ranked as Data Deficient.

Theoxena *sp.* 'non-pectinate'

■ Rarities & Specialities ■

Macarostola miniella ■ WS 11–13mm

DESCRIPTION Highly distinctive endemic moth in the Gracillariidae family, only found in the North Island. Adult moths have two colour variations. The most common is the crimson and yellow form (right). The other has more brownish colouration replacing the crimson (left). Unlikely to be confused with any other moth species. **DISTRIBUTION** Throughout the North Island, especially Taranaki, Waikato, Auckland and Northland regions. **HABITS AND HABITATS** Larvae are leaf miners on the endemic tree Swamp Maire *Syzygium maire*. Larvae roll the leaves of their host plant, both to feed from and then pupate in. Recorded throughout the year, but most commonly in early summer.

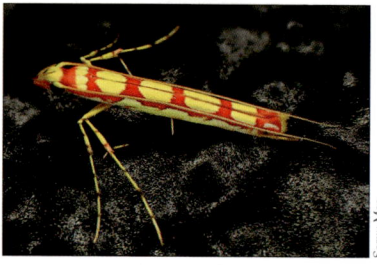

Green-toothed Owlet ■ *Ichneutica chlorodonta* WS 27–36mm

DESCRIPTION Endemic moth in the Noctuidae family. Very variable in marking and colouration on forewings, but always has very bright green or green-yellow colouration overlaying a brown or greyish background colour. Pale purplish edging to cross-lines also quite distinctive. Reniform and orbicular spots often bright green with brown or blackish centres. **DISTRIBUTION** Throughout the North, South and Rakiura/Stewart Islands in well-forested regions. Rare on the eastern South Island. **HABITS AND HABITATS** Associated with native forest and shrubland, but host plant/s and life history are unknown. Will come to light. Adults on the wing in September–April.

■ RARITIES & SPECIALITIES ■

Exquisite Olearia Owlet ■ *Meterana exquisita* WS 30–36mm

DESCRIPTION Exquisite beauty of *Olearia* shrubland. Colour pattern pale blue/green mixed with contrasting white and black markings in pattern form that resembles lichen and blends in well with its natural small-leaved *Olearia* shrubland habitat, as well as lichen-encrusted branches and rocks. **DISTRIBUTION** Reported in Auckland, Waikato, Taupo, Whanganui, Wairarapa, Nelson, South Canterbury, Mackenzie country, Central Otago, Otago Lakes and Southland; also recently on Stewart Island. However, some populations have become rare or extinct due to habitat loss. **HABITS AND HABITATS** Larval host species are small-leaved *Olearia* spp., such as *O. hectorii*, *O. odorata*, *O. lineata*, *O. fimbriata*, *O. solandri* and *O. bullata*. Only one generation a year. Larvae bright green and well camouflaged when feeding on their host species. They feed for one month before they pupate. Adults on the wing at night in August–December but most common in September–October. Attracted to light. **REMARKS** Now rare in some regions and has been extirpated from some sites, including its type locality in Southland. Classified as At Risk, Relict by the Department of Conservation. Loss of host plants due to land development has caused this decline. As such, protection, and expansion of shrubland containing small-leaved *Olearia* is important for their conservation, as well as that of other related moths such as the Grand Olearia Owlet (opposite).

▪ Rarities & Specialities ▪

Grand Olearia Owlet ▪ *Meterana grandiosa* WS 44–50mm 🅔

DESCRIPTION Tremendous, strongly marked endemic noctuid. Antennae and legs brownish-yellow. Forewings purplish-brown in base colour suffused with yellowish. Dark black markings on forewings highly distinctive and unique to the species. Larvae green in appearance with broad white lateral stripe. As they mature, they turn pinkish and can grow to 3.3cm in length. **DISTRIBUTION** Reported in the Wairarapa, Central Otago, Otago Lakes, Dunedin and Southland. However, has become locally extinct in some localities. Recent reports are from very limited areas concentrated in Central-Western Otago and between Palmerston North and Dannevirke. **HABITS AND HABITATS** Larval host plants are small-leaved *Olearia* species, such as *Olearia hectorii*, *O. fragrantissima*, *O. fimbriata* and *O. odorata*. Only one generation per year. Late-autumn to early winter emerging moth, being on the wing from mid-April to early June. Thus, astonishing in its ability to fly on cold, early winter Central Otago nights. **REMARKS** Potentially extinct in some former strongholds and rare in other areas where it still remains. Classified as At Risk, Relict by the Department of Conservation due to loss of its *Olearia* hosts, as described for the Exquisite Olearia Owlet (opposite). As such, protection, and expansion of shrubland containing relevant small-leaved *Olearia* species used by the larvae, such as *O. hectorii* and *O. odorata*, is crucial for their conservation. Appears to be rarer than the spring emerging Exquisite Olearia Owlet, as it has a more restricted distribution and does not use the reasonably widespread *O. bullata* as a host plant.

▪ Rarities & Specialities ▪

Southern Pimelea Owlet ▪ *Meterana meyricci* WS 38–42mm ⓔ

DESCRIPTION Vibrant endemic moth adorned with a splash of bright pink. Yellow-greenish markings and bright white reniform spot distinctive on otherwise dark-coloured forewings. Adults also have flush of pink on abdomen and hindwings that is very conspicuous when in flight. Unlikely, therefore, to be confused with any other Noctuidae in the country. **DISTRIBUTION** Widely distributed in montane and alpine areas of the South Island. Mostly associated with subalpine and alpine zones to at least 2,000m. Can be locally abundant. **HABITS AND HABITATS** Larval host species are *Pimelea* spp., such as *P. poppelwellii*. Adults mostly on the wing in January–March, with a few in November and December. **REMARKS** There are two other closely related *Pimelea*-feeding *Meterana* species, M. *pictula* and an undescribed species from coastal habitats in Southland and Stewart Island.

Patchwork Owlet ▪ *Meterana pauca* WS 32–40mm ⓔ

DESCRIPTION Another moth species in the endemic genus *Meterana*. Head and body brownish-black. Distinctive light blue or light green patches cover thorax and forewings, resulting in a unique patchwork pattern. **DISTRIBUTION** Widespread from central North Island south and records from throughout the South Island. **HABITS AND HABITATS** Appears to occupy damp, forested areas and shrubland, including beech forest, but host plant is unknown. On the wing in October–January.

■ Rarities & Specialities ■

Nola parvitis ■ WS 18–20mm (e)

DESCRIPTION Beautiful but subtle, creamy white moth. Aotearoa New Zealand's lone endemic nolid (from the global family Nolidae, or tufted moths). **DISTRIBUTION** Reasonably widespread in the South Island, especially Otago, but not commonly seen. Found right across Otago from the shores of Lake Wakatipu to Otago Peninsula. Also recorded near Te Anau and in Nelson province, as well as a few North Island records from Lake Waikaremoana. **HABITS AND HABITATS** Host plant is the shrub *Helichrysum lanceolatum*. Appears to stay close to areas of this plant that are growing within tussock grassland and shrubland, or on forest edges and clearings. **REMARKS** Planting of *H. lanceolatum* shrubs on farms, gardens and parks will help create habitat for this special moth.

Alpine Casemoth ■ *Orophora unicolor* WS 25–27mm (e)

DESCRIPTION Endemic bag moth of the Psychidae family. Case covered with layers of short (to 10mm) lengths of tussock, laid longitudinally and overlapping, so that it looks like a bundle of twigs. Cases are 30–40mm long. Only the male metamorphoses into a recognizable moth. Brown with a large scruffy or hairy thorax. Adult female never leaves the bag and has no wings. **DISTRIBUTION** Widespread in the South Island, especially on dry eastern side of the Southern Alps. Common in northern part of Central Otago. **HABITS AND HABITATS** Feeds on tussock and *Ozothamnus*. Adult males come to light in September–December. Can be locally abundant. **REMARKS** Larva starts building a bag as soon as it hatches, by making a loop of silk, then adding more and more loops. Bag maintains humidity inside to prevent larvae from drying out. Cases are remarkably strong and difficult to tear apart. The larva pupates inside the bag. Males emerge as hairy grey-black moths and take flight. Female looks like a soft grub and stays safely inside the bag, releasing pheromones to draw the flying males to her.

Bag

▪ Rarities & Specialities ▪

Izatha psychra ▪ WS 21mm (e)

DESCRIPTION Subtly beautiful, small moth in the Oecophoridae family, endemic to the inland South Island. Forewings elongated and whitish. Pale greyish colouration, and narrow forewings with black basal streak and three black dots forming road triangle in basal half of wing distinctive within *Izatha* genus. Black scape of antennae (base or first segment) contrasts noticeably with whitish head. **DISTRIBUTION** Has only been found in the southern half of the South Island. Has not been seen at its type locality of Porters Pass since the nineteenth century, and until recently was only known from one small population at the Pukaki Scientific Reserve near Lake Pukaki in the Mackenzie Basin. In August 2020, a large fire damaged the reserve, and the moth was for a time thought possibly extinct. However, a survey completed in 2021 caught two adult males in a small patch of unburnt habitat. More recently, in 2022/23, two new populations were located in Oteake Conservation Park (North Otago), extending the species' known range significantly to the south. **HABITS AND HABITATS** Like other *Izatha*, or lichen tuft moths, likely to feed on lichens or dead and decaying wood among shrubland. The two sites where it has been found in Oteake Conservation Park are near a stream, and are damp with abundant and diverse grey shrubland, including *Olearia bullata*, a range of small-leaved *Coprosma* species, *Corokia cotoneaster*, *Gaultheria*, Totara and Matagouri. At the Pukaki Scientific Reserve, the habitat is shrubland, also with much *Corokia*, *Coprosma* and Matagouri, but lacking *Olearia*. **REMARKS** Ranked as Threatened–Nationally Endangered by the Department of Conservation. However, the recent sites found in Oteake Conservation Park may indicate that it is more widespread, and less immediately imperilled, than previously thought.

■ Further Reading ■

The following is a brief list of excellent books and other resources on butterflies and moths that were all of great help in the preparation of this book. There is plenty to learn and explore about Aotearoa New Zealand Lepidoptera in these resources.

Online Resources
- **iNaturalist** https://inaturalist.nz/
- **The Entomological Society of New Zealand** https://ento.org.nz
- **The Moths and Butterflies of New Zealand Trust** www.nzbutterflies.org.nz
- **Online image gallery to large New Zealand moths** https://oldwww.landcareresearch.co.nz/resources/identification/animals/large-moths/image-gallery
- **PlantSynz, invertebrate herbivore biodiversity assessment tool** plant-synz.landcareresearch.co.nz

Books & Other Publications

Crowe, A. 2002. *Which New Zealand Insect?* Penguin Books.

Dugdale, J. S., Emmerson, A. W., & Hoare, R. J. B. 2023. *Declana* and *Ipana* (Insecta: Lepidoptera: Geometridae: Ennominae). *Fauna of New Zealand* **82**.

Gaskin, D. E. 1966. *The Butterflies and Common Moths of New Zealand*. Whitcombe and Tombs, Christchurch.

Gibbs, G. W. 1980. *New Zealand Butterflies, Identification and Natural History*. William Collins, Auckland.

Hoare, R. J. B. 2010. *Izatha* (Insecta: Lepidoptera: Gelechioidea: Oecophoridae) (PDF). *Fauna of New Zealand* 65: 1–201.

Hoare, R. & Ball, O. 2014. *A Photographic Guide to Moths and Butterflies of New Zealand*. New Holland Publishers (NZ) Ltd.

Hoare, R. J. B., Dugdale, J. S., Edwards, E. D., Gibbs, G. W., Patrick B. H., Hitchmough, R. A. & Rolfe J. R. 2017. Conservation status of New Zealand butterflies and moths (Lepidoptera), 2015 (PDF). Wellington, New Zealand: New Zealand Department of Conservation. p. 8.

Hoare, R. J. B. 2017. Noctuinae (Insecta: Lepidoptera: Noctuidae) part 1: *Austramathes, Cosmodes, Proteuxoa, Physetica*. *Fauna of New Zealand* 73: 1–130.

Hoare, R. J. B. 2019. Noctuinae (Insecta: Lepidoptera: Noctuidae) part 2: *Nivetica, Ichneutica*. *Fauna of New Zealand* 80: 1–455.

Hudson, G. V. 1928. *The Butterflies and Moths of New Zealand*. Wellington, Ferguson & Osborn Ltd.

Hudson, G. V. 1939. A supplement to the butterflies and moths of New Zealand. Wellington, Ferguson & Osborn Ltd. pp. 387–481, pl. 53–62.

Patrick, B. H. & Patrick H. J. H. 2012. *Butterflies of the South Pacific*. University of Otago Press, Dunedin.

Patrick, B. 2000. Lepidoptera of small-leaved divaricating *Olearia* in New Zealand and their conservation priority (PDF). Wellington, N.Z.: Department of Conservation, New Zealand.

Patrick, B. H. 2020. Winter emerging NZ moths. *Butterflies and Moths of New Zealand* 34: 9–10. Moths and Butterflies of New Zealand Trust.

Patrick, B. 2021. Indigenous host plants of New Zealand's endemic Lepidoptera. *Canterbury Botanical Society Journal* 52.

White, E. G. (with contributions from J. S. Dugdale, R. J. B. Hoare & B. H. Patrick). 2002. *New Zealand Grassland Moths: A Taxonomic and Ecological Handbook Based on Light-trapping Studies in Canterbury*. Manaaki Whenua Press, Lincoln.

INDEX

Achyra affinitalis 19
Admiral, Honshu
White 126
New Zealand Red 127
Yellow 128
Aenetus virescens 81
*Agonopterix
alstromeriana* 33
umbellana 33
Agrius convolvuli 135
Agrotis ipsilon 85
Alpine Grey, Greater 104
Amblyptilia falcatalis 132
repletalis 132
Anachloris subochraria 42
Angle-wing Moth, Small 149
Anisoplaca achyrota 40
Antiscopa epicomia 19
Aoraia rufivena 153
Apoctena conditana 139
flavescens 139
Aponotoreas anthracias 44
dissimilis 155
insignis 45
synclinalis 155
Apple Moth, Light Brown 145
Arctesthes catapyrrha 45
Argyrophenga spp. 121
antipodum 121
janitae 121
Armyworm, Northern 110
Southern 110
Sugar Cane 105
Armyworm Moth, Tropical 113
Arrowhead 21
Artona martini 151
Asaphodes abrogata 42
adonis 156
aegrota 42

beata 43
chlamydota 43
clarata 44
ida 156
prasinias 157
Austramathes purpurea 85
*Austrocidaria
callichlora* 46
cedrinodes 46
gobiata 47
similata 47

Bactra noteraula 139
Bag Moth, Australian Lichen 130
Common 131
Bamboo Moth 151
Barea spp. 114
exarcha 114
Bell Moth, Sharp-tipped 144
Bird Nest Moth 138
Bityla defigurata 86
Black Tabby, Eastern 21
Blotched Moth, Green 87
Blue, Common 119
Long-tailed 118
Southern 119
Blue Moon 124
Burnished Brass, Slender 113

Cabbage Butterfly 128
Cabbage Tree Moth 53
Capua intractana 140
'*Capua*' *semiferana* 140
Carpet, Barred
Coprosma 47
Barred Pink 75
Dark Coprosma 47
Dotted Green 43
Elegant 43
Exquisite 156
Golden Grass 42
Green Coprosma 46
Large Striped 44
Yellow and Brown 157

Carpet Moth, Dark-banded 54
Casemoth, Alpine 170
Catamacta gavisana 140
Chalastra aristarcha 48
pellurgata 48
'*Chloroclystis*' *filata* 49
Chrysodeixis eriosoma 86
Cinnabar Moth 39
Cladoxycanus minos 79
Cleora scriptaria 49
Clepsis divulsana 141
leucaniana 141
'*Cnephasia*' *jactatana* 141
comma-mark cutworms 112
copper butterflies 120
Copper, Common 120
Central Otago Boulder 120
Central Otago Copper 120
*Coscinoptycha
improbana* 17
Cosmodes elegans 87
Cotton Bollworm Moth 89
Cotton Tipworm Moth 142
Cranberry Moth, Native 64
Crocidosema plebejana 142
Cryptaspasma querula 142
Ctenoplusia limbirena 86
Ctenopseustis fraterna 143
obliquana 143
Culladia cuneiferellus 20
Cydia succedana 143

Danaus plexippus 122
Dasypodia cymatodes 37

Dasyuris callicrena 157
Dead Sheep Moth 138
Deana hybreasalis 20
Declana floccosa 51
nigrosparsa 158
niveata 50
Delicate, Red-spotted 50
Dialectica scalariella 78
Diamondback Moth 129
Diarsia intermixta 88
Diasemia grammalis 21
Dichromodes spp. 159
gypsotis 159
ida 159
simulans 159
*Diplopseustis
perieresalis* 21
Dipterina imbriferana 144
Dodonidia helmsii 123
*Dumbletonius
unimaculatus* 80

Elachista spp. 36
thallophora 36
Elvia glaucata 52
Endrosis sarcitrella 115
Epalxiphora axenana 144
Epicyme rubropunctaria 50
Epiphryne undosata 53
verriculata 53
Epiphyas postvittana 145
Epyaxa lucidata 55
rosearia 55
venipunctata 55
Erebiola butleri 126
Erechthias capnitis 137
terminella 137
Eudonia spp. 22
aspidota 22
cataxesta 22
diphtheralis 23
feredayi 23
leptalea 23
melanaegis 22

INDEX

octophora 23
philerga 22
sabulosella 23
steropaea 22
submarginalis 22
trivirgata 23
Eurythecta zelaea 145
Eutorna symmorpha 34
Evening Moth, Brown 54
  Lesser Brown 54

Featherfoot, Grey-lined 136
Feredayia graminosa 88
Fern Moth, Golden-brown 27
Flat Moth, Tarata 34
Flax Notcher Moth 103
Forest Flash, Brown 71
Fruit Borer, Orange 146
Poroporo 26

Gadira acerella 24
Gall Moth,
  Muehlenbeckia Stem 136
Gellonia dejectaria 54
  pannularia 54
Ghost Moth, Forest 80
  Otago 153
  Winter 79
Glaucocharis spp. 25
  auriscriptella 25
  elaina 25
  lepidella 25
  pyrosphanes 25
Glyphipterix spp. 78
  achlyoessa 8
  barbata 78
  triselena 78
Gorse Pod Moth 143
Gorse Soft Shoot Moth 33
Grass Moth, Common 28
  Mire 29
grass miner moths 36

Guava Moth,
  Australian 17
Gum Moth, Emperor 135
Gymnobathra spp. 148
  hyetodes 149
  parca 149
  sarcoxantha 148
  tholodella 149

Harmologa amplexana 146
Hawkmoth,
  Convolvulus 135
Helastia spp. 56
  alba 56
  christinae 56
  cinerearia 56
  corcularia 56
  cryptica 56
  expolita 56
Helicoverpa armigera 89
Heloxycanus patricki 154
Heterocrossa spp. 12
Hofmannophila pseudospretella 115
Homodotis falcata 57
  megaspilata 57
house moths 115
'*Hydriomena*' *deltoidata* 54
  rixata 58
Hygraula nitens 26
Hypolimnas bolina 124

Ichneutica agorastis 89
  arotis 90
  atristriga 90
  averilla 91
  ceraunias 91
  chlorodonta 166
  cuneata 92
  disjungens 92
  epiastra 93
  infensa 93
  insignis 94
  lignana 94
  lithias 95
  maya 95

moderata 96
mollis 96
morosa 97
mutans 97
nullifera 98
omoplaca 98
paracausta 99
plena 99
propria 100
purdii 100
rubescens 101
scutata 101
semivittata 102
skelloni 102
steropastis 103
sulcana 103
ustistriga 104
virescens 104
Ipana atronivea 59
  egregia 60
  glacialis 160
  griseata 59
  junctilinea 60
  leptomera 61
Ischalis fortinata 61
  gallaria 62
  variabilis 62
Isotenes miserana 146
Izatha spp. 150
  acmonias 150
  convulsella 150
  huttonii 150
  katadiktya 150
  psychra 171

Kiwaia lithodes 40
Kowhai Moth 32
Kowhai Seed Moth 136

Lampides boeticus 118
Leaf Miner, Echium 78
leafrollers 143, 147
  Blacklegged 147
Leafroller Moth,
  Lucerne 141
Lepidoscia heliochares 130
  protorna 130
Leucania stenographa 105

Leucinodes cordalis 26
Leuroperna sera 129
lichen tuft moths 150
Limenitis glorifica 126
Liothula omnivora 131
Looper, Apple 68
  Blue and Orange Rock 159
  Common Fern 73
  Common Forest 72
  Five Finger 76
  Forest 70
  Green Garden 86
  Hook-tip Fern 73
  Huarau 73
  Kānuka 68
  Kawakawa 49
  Lacebark 53
  Large Hebe 77
  New Zealand 55
  Oblique-waved Fern 62
  Pale Fern 48
  Pome 76
  Rusty Hebe 157
  Silver Fern 48
  Small Hooked-tip 57
  Spotted Mānuka 61
  Streaked Inanga 160
  Striped Fern 62
  Tarata 76
  Winter 77
  Wirerush 155
Lycaena spp. 120
Lyre Moth 140

Macrostola miniella 166
Magpie Moth/
  Mokarakara 38
Māhoe Stripper Moth 88
Meal Moth 133
'*Megacraspedus*' *calamogonus* 41
Metacrias erichrysa 152
Meterana alcyone 105
  coeleno 106
  diameta 106
  decorata 107

Index

dotata 107
exquisita 167
grandiosa 168
levis 108
meyricci 169
ochthistis 108
pauca 169
praesignis 109
stipata 109
Mnesarchaea paracosma 84
Mnesictena flavidalis 27
Monarch Butterfly 122
Monopis crocicapitella 138
  ethelella 138
Musotima nitidalis 27
Mythimna separata 110

New Zealand Cutworm 97
Noctuid, Blood-spotted 112
Nola parvitis 170
Notoreas elegans 161
  paradelpha 63
Nyctemera annulata 38

Opodiphthera eucalypti 135
Opogona comptella 137
  omoscopa 137
Orange, Alpine Grassland 45
Orange Peel Moth 88
Orocrambus spp. 28
  aethonellus 29
  augustipennis 29
  corruptus 29
  cyclopicus 29
  flexuosellus 28
  ramosellus 28
  vittelus 28
  vulgaris 28
Orophora unicolor 170
Orthoclydon chlorias 161
  praefectata 63
Owlet, Alpine Treasure 95

Black & White 99
Common Snowgrass 91
Desert 92
Exquisite Olearia 167
Grand Olearia 168
Green Carpet 99
Green-marked 94
Green-toothed 166
Kowhai 107
Large Grey 104
Lawyer 106
Mottled Brown 109
Patchwork 169
Pōhuehue 106
Small-eyed 85
Southern Pimelea 169
Owlet Moth, Slender 39

Painted Lady, Australian 124
Painted Wedge 147
Pantydia sparsa 37
Paranotoreas brephosata 64
  fulva 162
  zopyra 65
Patagoniodes farinaria 133
Pasiphila spp. 66
  bilineolata 67
  dryas 67
  fumipalpata 67
  halianthes 67
  inducata 66
  lichenodes 67
  lunata 67
  magnimaculata 67
  melochlora 67
  muscosata 67
  nereis 67
  punicea 67
  rubella 66
  sandycias 67
  sphragitis 66
  testulata 66
Percnodaimon sp. 125

Persectania aversa 110
Phaeosaces apocrypta 35
  coarctatella 35
Phrissogonus laticostata 68
Physetica spp. 111
  caerulea 111
  cucullina 111
  longstaffi 111
  phricias 111
  prionististis 111
Pieris rapae 128
Plantain Moth 74
Planotortrix spp. 147
  exessana 147
  notophaea 147
  octo 147
Plume, Araliad 132
Plume Moth, Hebe 132
Plutella xylostella 129
Poecilasthena pulchraria 64
  schistaria 68
Poison Hemlock Moth 33
Pond Moth 26
porina moths 82
Pasture 82
Sphagnum 154
Summer 82
Proternia philocapna 32
Proteuxoa comma 112
  tetronycha 112
Pseudocoremia dugdalei 162
  fenerata 69
  indistincta 69
  leucelaea 70
  lupinata 70
  productata 71
  rudisata 71
  suavis 72
Pterophorus innotatalis 132
  monospilalis 132
pug moths 66
  Australian 49
  Emerald 67
  Green Broom 67

Lawyer 52
Mottled Forest 67
Mountain 67
Pūriri Moth 81
Pyralis farinalis 133
Pyrgotis plagiatana 147
Pyroderces apparitella 18

Rhapsa scotosialis 39
Ringlet, Black Mountain 125
  Butler's 126
  Forest 123

Sarisa muriferata 73
Sarisophora leucoscia 83
Scar Bank Gem 87
Schrankia costaestrigalis 38
Scoparia, Boot 31
  Shining 23
Scoparia spp. 30
  exilis 30
  halopis 31
  minusculalis 31
  petrina 30
  rotuella 31
  subita 30
  ustimacula 31
Scopula rubraria 74
Scuttler, Dusky 137
sedge moths 78
Semilooper, Forest 51
Sestra flexata 73
  humeraria 73
Silvering, Yellow 25
Skeletonizer, Gum Leaf 83
Snout, Pinion-streaked 38
Snowberry Yellow 161
Speargrass Moth 98
spindle moths 163
Spindle, Kāmahi Green 74
  Tutu Green 163
Spodoptera litura 113
Stathmopoda spp. 136
  aposema 136

Index

plumbiflua 136
Stemborer, Tomato 41
Stem Borer 78
   Senecio Blue 133
Stericta carbonalis 134
Stone Jumper 40
Streak, Ruddy 114
Swan Plant Flower Moth 24
Sword Grass, Dark 85
Symmetrischema tangolias 41

Tachystola acroxantha 114
Tatosoma spp. 163
   agrionata 164
   alta 163
   fasciata 163
   lestevata 163
   monoviridisata 164
   topea 164
   transitaria 164
Tatosoma tipulata 74
Tebenna micalis 18
Theoxena scissaria 165

Theoxena sp. 'non-pectinate' 165
Thistle Moth, Small 18
Thysanoplusia orichalcea 113
Tiger Moth, Western 152
Tingena spp. 116
   actinias 116
   chloradelpha 116
   compsogramma 116
   phegophylla 116
Titoki Moth 134
Trachypepla spp. 117
   conspicuella 117
   contritella 117
   lichenodes 117
   protochlora 117
Triangle, Clematis 20
   Riverbed 45
   Rusty Dotted 27
Tussock, Common 121
   Janita's 121
tussock ringlets 121
Tyria jacobaeae 39

Underwing, Creekbed Orange 65
   Orange 64
Uraba lugens 83
Uresiphita maorialis 32

Vanessa gonerilla gonerilla 127
   itea 128
   kershawi 124
Vanicela disjunctella 134

Wainscot, Common Dotted 102
   Dark Underwing 103
   Orange Astelia 100
Wattle Moth, Northern 37
Web Spinner, Cotton 19
Window-maker, Flax 63
Wiseana spp. 82
   cervinata 82
   copularis 82

Xanthorhoe occulta 75
   semifissata 75
Xyridacma alectoraria 76
   ustaria 76
   veronicae 77

Yellow Dot 84
Yellow Field Bell Moth 141
Yellow Moth, Pink-tipped 148

Zealandopterix zonodoxa 84
Zebra Moth, North Island 59
   South Island 60
Zermizinga indocilisaria 77
Zizina otis 119
   oxleyi 119

Acknowledgements

I am very grateful for the encouragement and support I have received from some of New Zealand's most outstanding naturalists, entomologists and lepidopterists since I began to show an interest in moths and butterflies a mere few years ago. It has been a steep learning curve, but I have soaked up a plethora of new information in a short time thanks to your assistance. In particular, I have learnt a huge amount from discussion and interaction with Shaun Murphy, Neville Hudson, Dr Robert Hoare, Samuel Purdie and Brian Patrick. I would not have been able to write this book without this encouragement, as well as the knowledge I have garnered from these fellow moth and butterfly fanatics. It has been a very enjoyable and rewarding experience becoming a part of New Zealand's community of Lepidoptera enthusiasts and feeling that I am contributing positively to this field.

I am extremely grateful to Dr Robert Hoare for reviewing a draft of this book and providing key suggestions for improvements. I am also very thankful to Samuel Purdie for helping facilitate this opportunity and for providing some feedback on the draft. Thanks to all the photographers who allowed their beautiful images to be used in the book: Christopher Stephens, Saryu Mae, Shaun Murphy, Samuel Purdie, Neville Hudson, Sara Smerdon, Aalbert Rebergen, Helen Macky, Reino Grundling, Robert Hoare and Roger Frost.

Thanks to my wife and family for putting up with me talking endlessly about moths and disappearing on a whim to go and chase moths, only to arrive back at the house at 3 a.m. It is worth the many late nights.